D0765411

ROBOTICS FOR ENGINEERS

ROBOTICS FOR ENGINEERS

by Yoram Koren

Head, Robotics Laboratory
Technion—Israel Institute of Technology

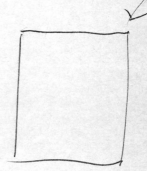

McGraw-Hill Book Company

New York   St. Louis   San Francisco   Auckland   Bogotá   Hamburg
Johannesburg   London   Madrid   Mexico   Montreal   New Delhi
Panama   Paris   São Paulo   Singapore   Sydney   Tokyo   Toronto

Library of Congress Cataloging in Publication Data
Koren, Yoram.
    Robotics for engineers.

    Includes index.
    1. Robotics.   2. Robots, Industrial.   I. Title.
    TJ211.K66   1985   629.8′92   84-21316
    ISBN 0-07-035399-9

 2345678910   VBVB   89876

ISBN 0-07-035399-9

The editors for this book were Betty Sun and David Fogarty, the
designer was M.R.P. Design, and the production supervisor was
Teresa F. Leaden. It was set in Caledonia by University Graphics,
Inc.

To my wife, Aliza,
my father, Shlomo,
and in loving memory of my mother, Bathia

Contents

Preface

Today's factories are designed either to mass produce identical products, or to manufacture a variety of products in limited numbers of "batches." Changing the manufacturing process is difficult and costly. Tomorrow's computer-based factories must adjust readily to fluctuations in market demand and to changes in designs, materials, and tooling. New roles are developing for robots and computers, and for the people who use them.

This revolutionary change in factory production techniques will require updating engineering education. There are four ingredients that universities need to educate the engineers required for modern manufacturing systems: (i) equipment, (ii) suitable curriculum, (iii) academic staff, and (iv) books to use in the instructional programs.

The present book was written for engineers who have to install or maintain industrial robots, and for students in mechanical and industrial engineering with interest in robotic systems and their implementation.

The book assumes no previous knowledge or experience in the area. The first two chapters of the book are introductory, and explain what industrial robots are, how they operate, what their main features are, and which tasks they can perform. The next three chapters (Chaps. 3, 4, and 5) deal with the design of control loops and kinematics. These three chapters are more mathematically oriented, and could be skipped if a thorough knowledge of the subject is not required. Chapters 6, 7, and 8 discuss various aspects of robotics: applications (Chap. 6), programming methods (Chap. 7), and sensors (Chap. 8). Chapter 9 presents a systematic approach to selecting robots for existing plants, a subject of extreme importance to unexperienced engineers in this field. The last chapter (Chap. 10) deals with the link between robotics and related fields, such as computer-aided design and manufacturing, and the integration of robots into flexible manufacturing systems. Chapter 10 also discusses future trends in the area: computer-integrated manufacturing systems and the factory of the future.

Each chapter in the book is a self-contained unit and is nearly independent of information provided in other chapters. For example, an engineer with interest in selecting a robot for a particular application might read only Chaps. 6 and 9 and use Appendix B which provides a directory of robot vendors. For convenience, I have included a glossary of robotics terms in Appendix A, and a list of robotics research institutes in Appendix C. I have tried to compile a complete list in Appendixes B and C, and apologize for any omissions.

ACKNOWLEDGMENTS

A major part of this book as well as the one preceding it, *Computer Control of Manufacturing Systems* (McGraw-Hill, 1983), were written at the University of Michigan in Ann Arbor, Michigan, during 1980–1982, when the author served as University of Michigan Visiting Goebel Professor of Mechanical Engineering at the College of Engineering, and Director of the Integrated Design and Manufacturing Division at the University of Michigan Center for Robotics and Integrated Manufacturing.

The author would like to express his gratitude to Professor James J. Duderstadt, Dean of the College of Engineering, and to Professor Daniel E. Atkins, Associate Dean for Research and Graduate Programs at the College and Director of the Center for Robotics and Integrated Manufacturing, for their generous support during his stay at the University. Their efforts to enhance the University of Michigan's programs in manufacturing and robotics are appreciated both by their colleagues and by leaders in industry.

The author is indebted to Professor J. G. Bollinger (Dean of the College of Engineering, the University of Wisconsin), who introduced him to the field of industrial robots some 11 years ago. Because the field of robotics is relatively new, there are a limited number of sources from which one is able to draw. The author was fortunate in learning much about robotics with his graduate students at the Technion, and he would like to thank them all, with special thanks to G. Amitai, J. Borenstein, O. Masory, and M. Shoham. The help of G. Amitai and M. Shoham in contributing two chapters (Chaps. 4 and 5) of this manuscript and the additional assistance of Dr. M. Zarrugh in reviewing these chapters are appreciated. The author would also like to thank Dr. A. Tevet (Towster) for reviewing major parts of the manuscript, Z. Kalmar for assisting with the drawings, and R. Rubinfeld for editing Table 9-1.

The author would like to express his thanks to the Palm Beach Chapter of the American Technion Society for supporting the Robotics Laboratory in the Faculty of Mechanical Engineering at Technion.

Finally, I thank my wife, Aliza, for her continued encouragement, and my children, Shlomik and Esther, for their patience throughout the time it took to complete this book.

YORAM KOREN

List of Abbreviations

ac = alternating current

AI = artificial intelligence

BRU = basic resolution unit

BBR = bend-bend-roll

CAD = computer-aided design

CAM = computer-aided manufacturing

CAT = computer-aided testing

CIM = computer-integrated manufacturing

CNC = computerized numerical control

CP = continuous path

CPU = central processing unit

CRT = cathode-ray tube

DAC = digital-to-analog converter

dc = direct current

DK = direct kinematics

FMS = flexible manufacturing system

GMAW = gas-metal arc welding

IK = inverse kinematics

JCS = joint coordinate system

MIG = metal inert gas

NC = numerical control

PTP = point-to-point

r = revolution

rad = radian

RBR = roll-bend-roll

rpm = revolution per minute

TCP = tool center point

TCS = tool coordinate system

TIG = tungsten inert gas

WCS = world coordinate system

CHAPTER 1

Basic Concepts in Robotics

Industrial robots are beginning now to revolutionize industry. These robots do not look or behave like human beings, but they do the work of humans. Robots are particularly useful in a wide variety of industrial applications, such as material handling, painting, welding, inspection, and assembly. Even more impressive, however, is the new perspective that robots may bring to the factory of the future. Current research efforts focus on creating a "smart" robot that can "see," "hear," "touch," and make decisions.

1.1 INTRODUCTION

Industrial robots, as other modern manufacturing systems, are advanced automation systems that utilize computers as an integral part of their control. Computers are now a vital part of industrial automation. They run production lines and control stand-alone manufacturing systems, such as various machine tools, welders, inspection systems, and laser-beam cutters. Even more sophisticated are the new robots that perform various operations in industrial plants and participate in full automation of factories.

   A revolutionary change in factory production techniques and management is predicted by the end of the twentieth century. Every operation in this factory of the future, from product design to manufacturing, assembly, and product inspection, would be monitored and controlled by computers and performed by industrial robots and intelligent systems. It is well to keep in mind that this automatically controlled factory is nothing more than a new phase in the industrial revolution that began in Europe two centuries ago and progressed through the following stages:

Construction of simple production machines and mechanizations were the first steps in this revolution that started in 1770.

Fixed automatic mechanisms and transfer lines for mass production came along as the second step at the turn of this century. The transfer line is an organization of manufacturing facilities for faster output and shorter production time. The cycle of operations is simple and fixed and is designed to produce a certain fixed product.

Next came machine tools with simple automatic control, such as plug-board controllers to perform a fixed sequence of operations, and copying machines in which a stylus moves on a master copy and simultaneously transmits command signals to servo drives.

The introduction of a new technology, numerical control (NC) of machine tools, in 1952, opened a new era in automation.

The logical extension of NC was computerized numerical control (CNC) for machine tools (1970), in which a minicomputer is included as an integral part of the control system.

Industrial robots have been developed simultaneously with CNC systems. The first commercial robot was manufactured in 1961, but they did not play a major role in manufacturing until the late 1970s.

The next logical extension of the two preceding steps is the fully automatic factory, which requires unprecedented involvement of computer-controlled systems and robots operating in concert in the production and assembly processes.

The new era of automation, which started with the introduction of NC machine tools, was undoubtedly stimulated by the digital computer. Digital technology and computers enabled the design of more flexible automation systems, namely systems which can be adapted by programming to produce or assemble a new product in a short time. Actually, *flexibility* is the key word which characterizes the new era in industrial automation. Robots and manufacturing systems are becoming more and more flexible with progress in computer technology and programming techniques.

1.2 ADVANTAGES AND APPLICATIONS OF ROBOTS

The term *robot* comes from Czech and means "forced labor." The term in its present interpretation was invented by the Czech writer Karel

Capek in his 1921 *R.U.R.*, which stands for Rossum's Universal Robots. Although Capek's robots look like people, they do not have human feelings, and they work twice as hard as human beings.

Industrial robots do not look like human beings but they do the work of humans. The concept of an industrial robot was patented in 1954 by G. C. Devol (U.S. Patent No. 2988237). Devol describes how to construct a controlled mechanical arm which can perform tasks in industry. The first industrial robot was installed by Unimation Inc. in 1961, and since then thousands of robots have been put to work in industry in the United States, Japan, and Europe.

The present industrial robots are actually mechanical handling devices that can be manipulated under computer control. A typical structure of a robot system is shown in Fig. 1-1. The mechanical handling device, or the manipulator, emulates one arm of a human being, and similarly has joints, denoted sometimes as shoulder, elbow, and wrist. The wrist contains pitch, yaw, and roll orientations. The joints are driven by electric, pneumatic, or hydraulic actuators, which give robots more potential power than humans.

The computer, which is an integral part of every modern robot system, contains a control program and a task program. The control program is provided by the robot manufacturer and enables the control of each joint of the robot manipulator. The task program is provided by the user and specifies the manipulator motions required to complete a specific job. Task programs are generated either by leading the robot through the required job or by using off-line programming languages. When a programming language is used, the robot computer also contains a language processor which interprets the task programs and provides the data required by the control program to direct the robot's motions. The control program uses the task program as data, and therefore, for every job a new task program must be generated by the user.

The industrial robot can do a human's work much more effectively. Robots work two shifts a day (three shifts in Japan), 8-hours per shift. They do not take breaks or go on strike, and they know neither weariness nor boredom. In Chrysler's Jefferson plant, 200 human welders on the assembly line were replaced by 50 robots.† These robots work two shifts, and the assembly line's output has increased by almost 20 percent.

American industry hopes that robots will provide an answer to one of its major problems: the decline in productivity. From 1947 to 1965, United States productivity increased by 3.4 percent a year. The growth rate decreased to 2.3 percent in the next decade, then dropped to below 1 percent in the late 1970s, and in 1980 the rate became nega-

†*Time*, December 8, 1980, p. 72.

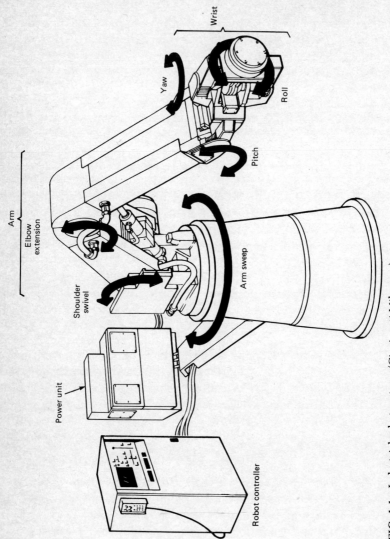

FIG. 1-1 Industrial robot system. (Cincinnati Milacron.)

4

tive, namely, a decline in productivity. In the same period, Japanese productivity increased at an average annual rate of about 7.3 percent. With the new trend in industry of incorporation of more robots and other computerized automation in the production lines, the nation's productivity will eventually increase in the 1980s.

Nevertheless, the better quality of products achieved by robot operations rather than the higher productivity, is sometimes regarded as the most important contribution of robots. If a robot is instructed to make 20 spot welds on a car, it will always make 20 welds, while a human operator might sometimes weld fewer points. When robots are operating on assembly lines in the automotive industry, terms such as "Monday car" do not exist. The product quality is more consistent and quality control is simpler.

The first robots replaced people in hazardous and dangerous tasks. A remote-controlled manipulator which was used in the NASA/AEC‡ Space Nuclear Propulsion Program was developed at Case University by H. W. Mergler in the mid sixties.† In nuclear plants today robots are transporting reprocessed plutonium, one of the most toxic substances known. Until the robot age, this dangerous task was done by workers wearing elaborate protective suits.

Industrial robots are ideally suited for jobs which are regarded as unpleasant or unhealthy for people. Spray painting, for example, is unhealthy to the operators who breathe toxic air. These operators would be happy to be moved to less hazardous assignments when spray painting robots replace them by emulating the operators' motions.

Robots are ideal for any repetitive job which is regarded as too boring or too tedious for human operators. In many assembly lines the production process consists of numerous simple assembly tasks, each of a limited number of operations which require unskilled labor. Robots can be easily integrated into such processes and do the job more effectively than people. Robots are, however, useless for any intelligent work in which the creative abilities of a human being are needed. In short, robots enable the use of people for what people do best.

The advantages of industrial robots can be summarized as follows:

1. Flexibility
2. High productivity
3. Better quality of products
4. Improved quality of human life by performing the undesirable jobs.

---

‡The Atomic Energy Commission has been succeeded by the Energy Research and Development Administration.
†*Life Magazine*, August 25, 1966.

Many industrial leaders perceive the robot only as a programmable machine that can replace workers. But the key issue is not putting a robot in a shop to do the work of two persons. It is the flexibility of the robot which enables its use as a tool in the factory of the future.

In the factory of the future, raw materials will be transformed into finished products completely by automation. Every operation in this future factory—from product design and manufacturing to assembly and production inspection—will be monitored and controlled by computers and performed by robots and artificial intelligence systems.

An important aspect of this factory of the future will be its manufacturing flexibility. By simply changing the computer program and an appropriate simple setup, one will be able to change the factory's finished product. The flexibility of the factory gives the company greater freedom to meet different specifications for a given product line and to introduce new models more quickly. A Japanese company which applied flexible production lines was able to ship out a new model color television less than half a year after receiving specifications from the marketing department.† In the factory of the future this time will be shorter, even for more complex products. Without the participation of robots in manufacturing, assembly, and inspection, the factory of the future cannot be realized.

Industrial robots are mainly used today in the following applications:

Loading and unloading of machines

Spot welding

Spray painting

Assembly

Arc welding

Inspection

Die-casting and forging operations

Drilling and deburring of metal parts

Robot use in the United States and Japan by the end of 1982 is broken down in Fig. 1-2. In the early eighties the portion of robots performing arc welding, assembly, and inspection has gradually increased. It is worthwhile to note that each of the robot applications requires different positional accuracy and different features of the control system.

†*The Wall Street Journal*, February 16, 1982.

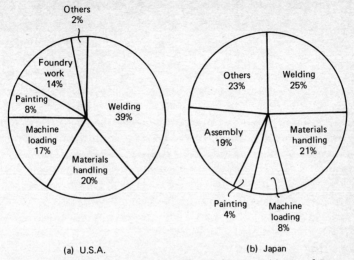

(a) U.S.A.   (b) Japan

FIG. 1-2 Breakdown of robot usage by applications. (*a*) United States; (*b*) Japan. (Data from P. Aron, "The Robot Scene in Japan," Report No. 26, Daiwa Securities America Inc., September 1983.)

1.3 NONINDUSTRIAL APPLICATIONS

So far we have discussed only industrial applications of robots. Nevertheless, robot systems can be used in other fields as well. Perhaps the most unusual to date is the use of a robot for shearing all of the wool from sheep. It is being done in Australia where an experimental robot developed in 1979 has been used in over 150 trials on live animals (Stauffer, 1982; Wong & Hudson, 1983). The shearing device is an eight-axis hydraulic arm controlled with a minicomputer using software developed specifically for the shearing operation. Electrical sensing techniques are used to detect the sheep skin, and sheep presentation equipment has been designed to hold and manipulate the animals into different positions for shearing.

Functional requirements of the shearing arm are divided into two sections. Upper and lower links mounted on a traverse carriage carry the shearing head and present it at the correct position and tilt angle. The cutting head has three angular degrees of freedom to maintain the correct cutter attitude. A follower actuator, maintained approximately normal to the sheep skin, controls the height of the cutter above the skin in response to sensor readings. Resistance and capacitance sensors mounted on the cutter head are used for sensing the skin.

Industrial robots are already being used for applying ribbons of adhesives or sealant on a variety of workpieces. Decorating chocolates with piped patterns is a generally similar task. In this application, a vision-equipped robot arm is used. The robot carries a piping nozzle and decorates the chocolates as they pass by at random locations on a conveyor belt. The robot adapts to the exact position and orientation of each chocolate and also recognizes pieces that vary in size or shape and must be discarded. The system uses a television camera to scan the chocolate at a point on the belt ahead of the robot position. Feature extraction routines are applied to the acquired data to extract key parameters about the candy shapes (Staufer, 1982). Future versions of the equipment would probably incorporate more than one arm per scanning system.

A different application of robots is in space technology. The highlight in the first flights of the space shuttle *Columbia* in 1981–1982† was the operation of the arm in space (see Fig. 1-3). The arm is a 50-ft mechanical manipulator controlled by the astronaut from a post at the rear of *Columbia's* cockpit. The astronaut looks at a target point with the aid of television cameras attached to the manipulator and tries to bring the end of the arm to this point. On the third flight of *Columbia* the arm was tested with a 353-lb payload. The task of this remote-controlled manipulator is to place satellites into orbit and retrieve them when they fail, which is an essential task in advancing space technology.

A similar robot system could be used in hospitals to help paralytic people or those who must be in bed after surgery. With the aid of a tiny joystick, a sick person could instruct the robot to bring medicine, to open the door, or turn on the television set. By adding a voice communication system, the robot could be taught to respond to voice instructions of the sick person, thus relieving nurses of nonprofessional tasks.

The household robot is another dream. Every housekeeper wants some help in performing household tasks: cleaning the house, washing the dishes, making the beds, taking out the garbage, etc. Unfortunately, these tasks cannot be performed by the robots of the early eighties. Today's robots are one-hand devices without vision and tactile sensors, and therefore they are appropriate only for routine and simple tasks. The housekeeping tasks, however, are not as routine as one might think. If you want to estimate whether a robot can do a specific job, you should (1) close your eyes, (2) use one hand only, and (3) put on a mitten. The mitten neutralizes your fingers' sense of touch and limits

---

†*Time*, "Putting an Arm on Space," Nov. 9, 1981, p. 81.

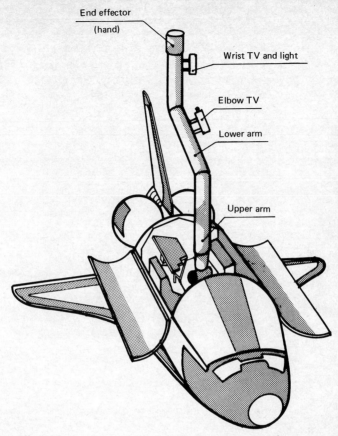

End effector
(hand)

Wrist TV and light

Elbow TV

Lower arm

Upper arm

FIG. 1-3   The space shuttle *Columbia* with the manipulator arm.

the operation of your hand to that of a two-finger gripper (without tactile sensor). Now, under these restrictions, try to take dishes off the table, pick up a chair and move it, or wash a window. It is not easy! Likewise, the same test can also be applied when considering a new robot installation in a factory.

What will house robots look like? They might look like R2-D2 or C3-PO, the *Star Wars* robots (Lucas, 1976). R2-D2 is a meter-high robot with two short legs and a cylindrical face containing a single radar eye. R2-D2 can only communicate with another robot in a series of electronic sounds. C3-PO is a human-size robot with a humanlike metallic face. C3-PO is a human–robot relations specialist and can translate the electronic language of robots to human language.

Another possibility is that the house robot will look like the one in

FIG. 1-4 A nonindustrial robot. (Demonstrated by Nikon in Robot VI, Detroit, Michigan, March 1982.)

Fig. 1-4. This robot is mobile, using a four-wheel drive mechanism (three-wheel drive might also be used), has a small television in its chest, and is equipped with two articulated arms. A video-tape recorder, a television camera, and a voice communication system could be added as well. At present such a robot has more entertainment value than practical use. It can greet guests or prepare cocktails and serve them to the guests. In the future, however, with the increased capability of computers and artificial intelligence techniques, these robots will be able to do much of the housekeeping. The robot arm will be able to use standardized attachments ranging from a vacuum cleaner to squeegees for washing windows, and will change the attachments by itself.

One of the most difficult problems with the house robot will probably be its inability to climb stairs, which will limit the variety of tasks that it could fulfill. Perhaps future houses will be designed to accommodate the house robot, which will be the servant of the house. The house will have no stairways or sharp corners and have rooms that are similar in shape to facilitate robot tasks. Eventually, we might have a bilevel society: elite human beings served by computerized robots.

1.4 BASIC STRUCTURE OF ROBOTS

The industrial robot is a programmable mechanical manipulator, capable of moving along several directions, equipped at its end with a work device called the *end effector* (or tool) and capable of performing factory work ordinarily done by human beings. The term *robot* is used for a manipulator that has a built-in control system and is capable of stand-alone operation.

One popular dictionary† defines a robot as "any mechanical device operated automatically to perform in a seemingly human way." By this definition, a garage door opener, which automatically opens the door by remote control, could also be a robot. Obviously this is not an industrial robot. The Robotics International Division of the Society of Manufacturing Engineers (RI/SME) defines the industrial robot as "a reprogrammable multi-functional manipulator designed to move material, parts, tools, or specialized devices through variable programmed motions for the performance of a variety of tasks." The key words in this definition are *reprogrammable* (can be programmed again and again), *multifunctional* (able to perform more than one special duty, but sometimes one), and *move material, parts, tools* (defines the task of the manipulator). By this definition, Mergler's manipulator (see Sec. 1.2) and the *Columbia* arm (see Sec. 1.3) are not robots since they are not programmable and operate on a master-slave basis. These devices are sometimes denoted as teleoperators, since they allow an operator to perform a task at a distance. However, the RI/SME definition of an industrial robot should include also the following key words: *motion along several directions* (or degrees of freedom), *end effector*, and *factory work*. If one of these key words is missing, then washing machines, automatic tool changers, or manufacturing machines for mass production might be defined as robots as well, and this is not our intention.

Modern robotic systems consist of at least three major parts:

1. The manipulator, which is the mechanical moving structure
2. The drives to actuate the joints of the manipulator
3. The computer as a controller and storer of task programs

In general, the structure of a robot manipulator is composed of a *main frame* and a *wrist* with a tool at its end. The tool can be a welding head, a spray gun, a machining tool, or a gripper containing open-shut jaws, depending upon the specific application of the robot. Each of

†With permission. From *Webster's New World Dictionary*, Second College Edition. Copyright © 1982 by Simon & Schuster, Inc.

these devices is mounted at the end of the robot and therefore is called the robot's end effector. The main frame is frequently denoted as the *arm,* and it typically consists of a sequence of mechanical links connected by joints. Each link is connected by a joint to the next link as shown in Fig. 1-5. The function of the joints is to control the motion between the links. The most distal group of joints affecting the motion of the gripper is referred to collectively as the wrist. A typical wrist included three joints which provide motions of roll, pitch, and yaw, as shown in Fig. 1-1. Each of the joints of both the arm and the wrist provides for one degree of freedom of motion of the robot end effector. Thus, an *n*-degrees-of-freedom manipulator contains *n* joints, or in more general terms, *n* axes of motion.

The motion of the end effector is generated by controlling the position and velocity of the robot's axes of motion. An axis of motion in robotics means a degree of freedom in which the robot can move. Basically the robot needs six axes of motion (or degrees of freedom) to reach an arbitrary point with a specific orientation in space. A different orientation might completely change the position of the robot arm. For example, to place a weld on the top side of the beam in Fig. 1-6 requires completely different orientation from that required to place a weld at almost the same point but on the bottom side of the beam, and consequently the position of the arm is changed. Typically the arm has three degrees of freedom, in linear or rotary motions, and the wrist section contains three rotary motions. The combination of these six motions will orient the robot's end effector and position it at the required point in space. Nevertheless, there are robot applications which require only five or four degrees of freedom. In these cases the wrist section contains less than three axes of motion.

The axes of motion of the robot arm can be either rotary or linear. A rotary axis is denoted in kinematics as a *revolute pair,* which is a simple

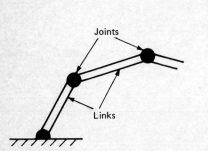

FIG. 1-5 Schematic structure of a robot arm.

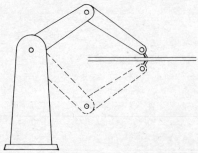

FIG. 1-6 Different robot orientations in beam welding.

FIG. 1-7   A nonservo robot. (Copperweld Robotics.)

hinge without axial sliding. A rotary axis is usually driven by an electric motor, which is coupled to the axis either directly or through a chain or a gear system. A linear motion of an axis can be obtained by a *prismatic pair* or by a leadscrew operation. In a prismatic pair a slider moves along one axis, in both positive and negative directions, such as the piston of a pneumatic or hydraulic actuator. The leadscrew transmits the motion of a rotary shaft to a linear displacement and is used when a rotary electric motor drives a linear axis in the robot arm.

Many processes require a robot capable of performing simple motions, such as picking up an oriented part and placing it into a machine. To perform the job, the robot requires only a limited number of sequential actions rather than sophisticated computer control of complicated motions. These simple robots have relatively primitive control strategies and are often referred to as *pick-and-place, limited-sequence,* or *nonservo* robots. A robot of this type is shown in Fig. 1-7. The drive elements of these robots are either pneumatic or hydraulic cylinders. The feedback devices are simply pairs of adjustable limit switches and mechanical stoppers. Two pairs are provided for each axis of motion of the arm to specify the end positions of the axis. The principle of operation of these robots is similar to that of automatic lathes with sequential control. At each step the controller of the robot sends a control signal to the valve of a desired axis to move it. The motion continues until restrained by the stopper and its corresponding limit switch. The limit switch sends a signal to the controller, which then commands the valve to close and proceeds to the next step in the control sequence involving another valve. This process is repeated until the entire sequence of steps has been executed.

Although the pick-and-place robots are relatively inexpensive and have satisfactory positional accuracy (better than 0.5 mm), they have a very low control flexibility, since the number of motions in a program is limited by the number of limit switches that can be installed. Further discussion will focus on the more advanced robots that have a high programming and control flexibility. These robots are often referred to as *servo-controlled* robots because they are controlled by closed loops with the aid of servomotor drives.

The new generation of robots refers to more intelligent robots which are capable of making decisions and generating "spontaneous" motions in response to novel situations. The decision making is based upon the information sent by sensors, such as a television camera for vision, or a force and pressure transducer for force feedback. The additional sensors are extremely useful in nonindustrial robot systems.

1.5 NUMERICAL CONTROL OF MACHINE TOOLS

NC machine tool systems are somewhat related to robotic systems. In both, the axes of a mechanical device are controlled to guide a tool which performs a manufacturing process task as shown in Fig. 1-8. In NC, the process may be drilling, milling, grinding, welding, etc., and in robotics may be painting, welding, assembly, handling, etc. In NC, the mechanical handling device is the machine tool, and in robotics it is the manipulator. In both systems each axis of motion is equipped with a separate driving device and a separate control loop. The driving device may be a dc servomotor, a hydraulic actuator, or a stepping

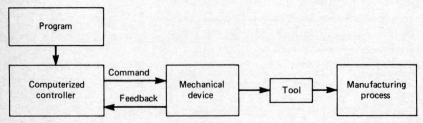

FIG. 1-8   Diagram of manufacturing systems.

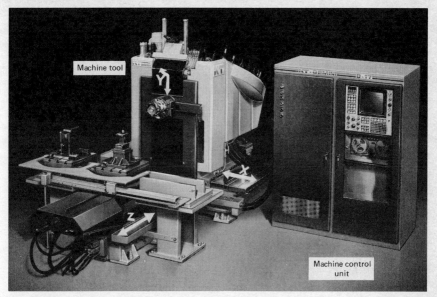

FIG. 1-9   CNC machine tool system—a machining center. (Kearney & Trecker.)

motor. The type selected is determined mainly by the power requirements of the machine or the robot.

The combination of the individual axes of motion generates, in both systems, the required path of the tool. In NC, the tool is the cutting tool, such as a milling cutter or a drill; in robotics, the tool is the instrument at the far end of the manipulator, which might be a gripper, a welding gun, or a paint-spraying gun. An axis of motion in NC means an axis in which the cutting tool moves relative to the workpiece. This movement is achieved by the motion of the machine tool slides. The main three axes of motion are referred to in NC as X, Y, and Z axes, as shown in Fig. 1-9.

In a typical NC system the numerical data which are required for producing a part are maintained on a punched tape and are called the *part program.* The part program in NC plays the same role that the task program plays in robotics. The part program is arranged in the form of blocks of information, where each block contains the numerical data required to produce one segment of the workpiece. The block contains, in coded form, all the information needed for processing a segment of the workpiece: the segment length, its cutting speed, feedrate, etc. Dimensional information (length, width, and radii of circles) and the contour form (linear, circular, or other) are taken from an engineering drawing. Dimensions are given separately for each axis of motion (X, Y, etc.). Cutting speed, feedrate, and auxiliary functions (coolant on and off, spindle direction, clamp, gear changes, etc.) are programmed according to surface finish and tolerance requirements.

Compared with a conventional machine tool, the NC system replaces the manual actions of the operator. In conventional machining a part is produced by moving a cutting tool along a workpiece by means of handwheels, which are guided by an operator. Contour cuttings are performed by an expert operator by sight. On the other hand, the operator of NC machine tools need not be a skilled machinist. The NC operator only has to monitor the operation of the machine, operate the tape reader, and usually replace the workpiece. All thinking operations that were formerly done by the operator are now contained in the part program. However, since the operator works with a sophisticated and expensive system, intelligence, clear thinking, and good judgment are essential qualifications of a good NC operator.

Preparing the part program for an NC machine tool requires a part programmer. The part programmer must possess knowledge and experience in mechanical engineering fields. Knowledge of tools, cutting fluids, fixture design techniques, use of machinability data, and process engineering are all of considerable importance. The part programmer must be familiar with the function of NC machine tools and machining processes and has to decide on the optimal sequence of operations. The part programmer writes the part program manually or by using a computer-assisted language, such as APT.

The controllers of modern NC systems include a dedicated mini- or microcomputer which performs the functions of data processing and control. These systems are referred to as *computerized numerical control* (CNC) systems and include hardware similar to that of robot systems. Robot systems, however, are more complicated than machine tool CNC systems for the following reasons:

1. Machine tools require the control of the position of the cutting edge of the tool in space. In principle, the control of three axes is adequate. Robots require the control of both the position of the tool center point (TCP) and the orientation of the tool, which is achieved by controlling six axes of motion.

The TCP is a point that lies along the last wrist axis at a user-specified distance from the wrist. It can be, for example, the edge of a welding gun or the center of a gripped object. In robotics the TCP plays the same role that the cutting edge plays in machining.

2. Continuous-path (CP) manufacturing systems (e.g., milling machines, arc-welding robots, etc.) require interpolators to determine

RESOLUTION, ACCURACY, AND REPEATABILITY   17

the path between given end points of segments. Axes of motion in machine tools are perpendicular to each other, thus creating a cartesian coordinate system (CCS). Motion along a straight path in CCS employs relatively simple linear interpolators (Koren, 1983) based on Eqs. (2-1) through (2-3). By contrast, most robots include rotary axes, but end points are usually given in a CCS whose origin is at the robot's base. Therefore machine tool interpolation methods cannot be applied in robotics.

Interpolation methods for CP robots are much more complex. The general approach consists first of breaking down the path between the end points into small sections along the same straight line (see Fig. 2-6). The motion from the beginning to the end of each section is then obtained by solving the inverse kinematic problem, or, in other words, transforming the points' cartesian coordinates to corresponding joint commands.

3. The structure of the robot manipulator is less rigid than that of a machine tool, and therefore, it is more difficult to achieve precise motions with a given level of accuracy in robotics.

4. The axes of motions in many robot systems are coupled (especially in the articulated structure, see Sec. 2.3.4). This means that a load on one axis affects the position accuracy of another axis. The effect of this coupling in machine tools is negligible.

5. NC and CNC machine tools use off-line part programming methods, which can be either manual or computer-assisted, such as the APT language. During off-line programming the machine remains in operation while the new part program is generated. By contrast, with most robot systems the robot itself is used during the teaching or programming stage (see Chap. 7).

6. In CNC machine tools and some robot systems, the velocity and position of the axes are controlled. In other robot systems, the torque and position are controlled. The control of the torque requires the use of the arm model in the control program which consequently complicates the control algorithm. Further details are given in Chap. 3.

1.6 RESOLUTION, ACCURACY, AND REPEATABILITY

The term *accuracy* in robotics is often confused with the terms *resolution* and *repeatability*. The resolution of a robot is a feature determined by the design of the control unit and is mainly dependent on the position feedback sensor. One has to distinguish between programming resolution and control resolution. The programming resolution is the smallest allowable position increment in robot programs and is referred to as the *basic resolution unit* (BRU), which might be on the order of 0.01 in (0.25 mm) in a typical linear axis, or 0.1° in a rotary axis. The control resolution is the smallest change in position that the feedback device can sense. For example, assume that an optical encoder which emits 1000 pulses per revolution of the shaft is directly attached to a rotary axis. This encoder will emit one pulse for each of 0.36° (360°/ 1000) of angular displacement of the shaft. The unit 0.36° is the control resolution of this axis of motion. Angular increments smaller than 0.36° cannot be detected. Best performance is obtained when programming resolution is equal to control resolution. In this case both resolutions can be replaced with one term: the system resolution.

The final accuracy of a robotic system depends on its mechanical inaccuracies, the computer control algorithms, and the system resolution. The mechanical inaccuracies are caused mainly by backlash in the manipulator joints and bending of the links. The backlash exists in gear mechanisms, in leadscrews, and in actuators of hydraulic drives. The minimization of the link bendings is the main design requirement for the link, as any deflection of the link due to the load at the robot's end causes positional errors. A higher rigidity of the links, however, should not be achieved by a substantial increase in their mass. A larger mass causes an increase in the time response of the arm, which consequently limits the operating speed of the robot. Control algorithms might cause position errors due to round-off errors in the computer. This problem, however, is usually taken care of by the control designer. System inaccuracy due to resolution is usually considered to be ½ BRU. The reason is that displacements smaller than 1 BRU can be neither programmed nor measured and, on the average, they count for ½ BRU. When the inaccuracies associated with the mechanical structure are included, a poorer accuracy will result. Therefore, the following relationship can be used to determine a realistic system accuracy:

$$\text{Robot accuracy} = \tfrac{1}{2}\,\text{BRU} + \text{mechanical accuracy}$$

Ideally, the accumulated effect of all mechanical inaccuracies should be under ½ BRU. This ensures that the robot accuracy is equal to the robot resolution. This is a reasonable demand when a cartesian coordinate robot is used, but it is hard to achieve with an articulated robot (see Sec. 2.3).

Repeatability is a statistical term associated with accuracy. If a robot joint is instructed to move by the same angle from a certain point a number of times, all with equal environmental conditions, it will be found that the resultant motions lead to differing displacements, as

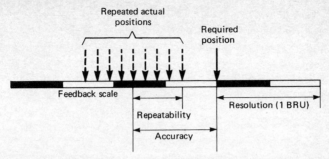

FIG. 1-10 Representation of resolution, accuracy, and repeatability of a robot arm.

shown in Fig. 1-10. System repeatability is the positional deviation from the average of these displacements. For example, ± 0.2 mm indicates that any point might be as much as 0.2 mm beyond or short of the center of the repeatability pattern. It is clear that through a greater number of such experiments we will be able to give a more precise estimate of the repeatability. The repeatability of robots will usually be better than the accuracy.

Most robot manufacturers provide a numerical value for the repeatability rather than the accuracy of their robots. The reason is that the accuracy depends upon the particular load that the gripper carries. A heavier weight causes larger deflections of the robot links and larger load on the joints, which degrades the accuracy. The repeatability value, however, is almost independent of the gripper load and therefore can be specified for any robot arm.

1.7 POSITION REPRESENTATION

The robotic systems which were produced through the sixties used electronic hardware based upon digital circuit technology. The robotic systems which were introduced in the seventies and eighties employ a mini- or microcomputer to control the robot axes and eliminate, as much as possible, additional hardware circuits. The computer accepts data in the form of task programs which describe the specific task of the robot, and the computer is able to process the data and provide command signals to actuators which drive the robot axes.

The required position of each axis of motion is expressed in both the hardware-based and computerized controllers by integers. Each integer unit corresponds to the position resolution of the axis of motion, i.e., the BRU. The BRU is also known as the *increment size* or *bit weight,* and in practice it corresponds approximately to the repeatability of the corresponding axis of the robot.

The digital controller in a hardware-based robotic system employs command pulses, where each pulse causes a motion of 1 BRU in the corresponding axis. Computerized robot systems can be designed in different configurations. In the simplest one, known as the *reference-pulse approach* (see Chap. 3), the computer transmits command pulses as well. These pulses can drive stepping motors in open-loop control, but usually they actuate dc servomotors or hydraulic actuators in closed-loop control. In closed-loop robot systems the position feedback device, e.g., the encoder, generates one pulse for a motion of each BRU. In all these cases a pulse is equivalent to displacement of 1 BRU:

$$\text{Pulse} \equiv \text{BRU} \tag{1-1}$$

The number of pulses transmitted to each axis or received from its encoder is equal to the corresponding axial position in BRUs, and the frequency of these pulses represents the axis velocity.

In the computer the information is arranged, manipulated, and stored in the form of binary words. Each word consists of a fixed number of bits, the most popular being the 8-bit and 16-bit words. The position of each axis is represented by integers with BRUs as units. For example, if the axial resolution is BRU = 0.05°, an angular displacement of 3.05° is given as 61 (3.05/0.05). This number is presented in an 8-bit microcomputer as 00111101, as shown in Table 1-1. Therefore in the control computer of the robot, each bit (*bi*nary digi*t*) represents 1 BRU:

$$\text{Bit} \equiv \text{BRU} \tag{1-2}$$

As can be seen from Table 1-1, when a practical BRU is applied, the 8-bit word limits the maximum axial position which can be instructed to the robot arm. Therefore, in practice, robotic systems use computers with a 16-bit word (or a double 8-bit word), which can represent up to 2^{16} = 65,536 different axial positions (including zero). If the axis resolution is, for example, BRU = 0.1 mm, this number represents

TABLE 1-1 Examples of Position Representation in 8-Bit Computer with BRU = 0.05°

Angular position, degrees	Position, BRUs	Binary representation
1.00	20	00010100
3.05	61	00111101
12.75	255	11111111

motions up to 6553.6 mm, or ± 3276.8 mm, and if the resolution is BRU = 0.005° the maximum angular displacement is 65536×0.005 = 327.68°.

Expressins (1-1) and (1-2) can be combined:

$$\text{Bit} \equiv \text{pulse} \equiv \text{BRU} \tag{1-3}$$

In many computerized robot systems, binary words rather than pulses are transmitted as the computer output. However, in these systems the actual position of the axis is measured by a digital feedback device (e.g., an incremental encoder) which transmits pulses expressing motion in BRUs. Therefore, actually in all types of computerized robot systems the terms *bit, pulse,* and *BRU* become equivalent and are interchangeably used in this text.

REFERENCES

Asimov, I.: *Robot,* Doubleday & Co., New York, 1950.

Capek, K.: *R.U.R. (Rossum's Universal Robots),* Doubleday & Co., New York, 1923.

Heer, E.: "Robots and Manipulators," *Mech. Eng.,* November 1981, pp. 42–49.

Koren, Y.: *Computer Control of Manufacturing Systems,* McGraw-Hill Book Co., New York, 1983.

Lucas, G.: *Star Wars,* Ballantine Books, New York, 1976.

Smith, D. N., and A. C. Wilson: *Industrial Robots: A Delphi Forecast of Markets and Technology,* SME and The University of Michigan, 1982.

Stauffer R. N.: "A Report on the 12th ISIR in Paris," *Robot. Today,* August 1982, p. 37–42.

Vonnegut, K.: *Player Piano,* Dell Publishing Co., New York, 1952.

Weisel, W. K.: "Industry, Research, Government and Robots—Are They Really Partners in Productivity?" *Commline,* January 1982, pp. 16–17.

Wong, P. C., and P. R. W. Hudson: The Australian Robotic Sheep Shearing Research And Development Programme, *13 Int. Symp. Ind. Robots & Robots 7,* April 1983, pp. 10–56.

CHAPTER 2

Classification and Structure of Robotic Systems

The classification of robotic systems can be done in three ways: (1) according to the type of system: point-to-point versus continuous-path; (2) according to the type of control loops: open-loop versus closed-loop; or (3) according to the structure of the manipulator: cartesian, cylindrical, spherical, or articulated. The selection of the type of system, control loop, and manipulator depends on the particular application. In addition, an appropriate wrist and end effector should be selected to fit the required application.

2.1 POINT-TO-POINT AND CONTINUOUS-PATH SYSTEMS

An individual with no experience in robots might think that a spot-welding robot and an arc-welding robot are the same, but with different welding equipment. Actually, these are two different robotic systems; the control of a spot-welding robot is based upon a point-to-point operation and is not recommended for performing arc welding, which requires a continuous-path system.

2.1.1 Point-to-Point Robotic Systems

A typical point-to-point (PTP) system is encountered in a spot-welding robot (see Sec. 6.2.1). In a spot-welding operation, the robot moves until the point to be welded is exactly between the two electrodes of the welding gun, and then the weld is applied. The robot then moves to a new point and another spot weld is performed. This process is repeated until all the required points on the part are welded. The welding gun is then brought to the starting point, and the system is ready for the next part.

In more general terms the description of a PTP operation is the following: The robot moves to a numerically defined location and then the motion is stopped. The end effector performs the required task with the robot stationary. Upon completion of the task, the robot moves to the next point and the cycle is repeated.

In a PTP system, the path of the robot and its velocity while traveling from one point to the next are without any significance. Therefore, a basic PTP system requires only axial position counters for controlling the final position of the robot tool in order to bring it to the target point. The coordinate values for each desired position are loaded into the counters with a resolution which depends on the system's BRU. During the motion of the arm the encoder at each joint transmits pulses which represent the position of the joint. Each axis of motion is equipped with a counter to which the corresponding encoder pulses are transmitted. At the beginning of a motion from a point, each axial counter is loaded by the corresponding required axial incremental distance to the next point in BRUs (see Sec. 1.6). During the motion of the arm, the contents of each counter are gradually decremented by the pulses arriving from the corresponding encoder. When all counters are at zero the robot is in its new desired position.

There are two basic structures of PTP robot systems. In the first one each axis moves from one point to the next as fast as it can, and therefore the path from the starting point to the end point is not controlled. This is illustrated in Fig. 2-1 for a motion in the plane of a cartesian coordinate robot (see Sec. 2.3.1).

The other structure applies to a more sophisticated PTP system in which the motion in all axes is terminated simultaneously. With this system, in each incremental motion the time required for the longest axial motion is calculated

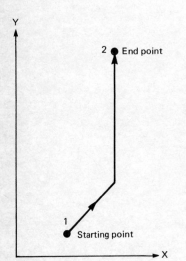

FIG. 2-1 Trajectory in a PTP cartesian robot system.

and used to determine the velocities of the other axes. The resulting trajectory is completely arbitrary, as illustrated in Fig. 2-2 for a cylindrical coordinate robot (see Sec. 2.3.2). However, since the time of motion t is identical in all axes, they arrive at the end point simultaneously, as shown in Fig. 2-2.

PTP robots are used in spot welding, material handling, loading and unloading of machines, and simple assembly tasks.

2.1.2 Continuous-Path Robotic Systems

In continuous-path (CP) robots, the tool performs the task while the axes of motion are moving, as in arc welding (see Sec. 6.2.2). The task of the robot in arc welding is to guide the welding gun along the preprogrammed path. In CP robots, all axes of motion may move simultaneously, each at a different velocity. These velocities, however, are coordinated under computer control in order to trace the required path or trajectory.

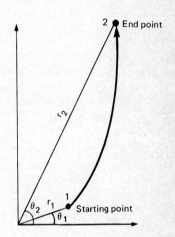

In CP operation, the position of the robot's tool at the end of each segment, together with the ratio of axes' velocities, determines the generated trajectory (e.g., the weld path in arc welding), and at the same time the resultant velocity also affects the quality of the work. For example, variations in the velocity of the welding gun in arc welding result in a nonuniform weld seam thickness (i.e., an unnecessary metal buildup or even holes).

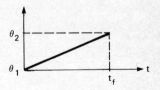

In CP cartesian robots (see Sec. 2.3.1), if the wrist has to trace a straight path of l length units in a velocity V, the axial velocities are

$$V_x = (x/l)V \qquad (2\text{-}1)$$

$$V_y = (y/l)V \qquad (2\text{-}2)$$

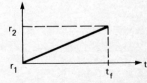

FIG. 2-2 Trajectory in a PTP cylindrical robot in which all axial motions are terminated simultaneously.

and
$$V_z = (z/l)V \qquad\qquad (2\text{-}3)$$
where
$$l = \sqrt{x^2 + y^2 + z^2}$$

and $x, y,$ and z are the components of l in the X, Y, and Z directions, respectively.

In CP robots a velocity error in one axis causes a path position error. For example, assume that the wrist of a cartesian robot has to move along a straight path in a 45° direction. If the velocity along the X axis is increased, the wrist moves in an angle smaller than 45°, as shown in Fig. 2.3. To avoid such errors, the system has to contain continuous-position control loops in addition to the position counters which check the end positions of the axes at each movement. Each axis of motion is equipped with a separate position control loop and a position counter. Required positions of the arm are specified separately to the appropriate position counter of each axis. The count in each counter is decremented by feedback pulses from the corresponding encoder until the desired position is reached (see Sec. 2.2).

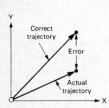

FIG. 2-3 Path error in CP cartesian system.

CP robots are used in arc welding, spray painting, deburring of metal parts, complicated assembly operations, and inspection.

2.1.3 Trajectory Planning

The task of the robot manipulator is defined by a sequence of points which are denoted as *end points* and stored in the robot computer. In a PTP robot the end effector is moved in an arbitrary path from one end point to the next. The end points in PTP robots are usually presented in joint coordinates, meaning the required position values of each axis are separately given and sent as the commands to the corresponding control loop.

A schematic diagram of a PTP articulated robot is shown in Fig. 2-4. The control program of the robot performs the trajectory planning, including acceleration, deceleration, and sometimes also time synchronization among the axes of motion. The program uses as data the joint coordinate values of the end points, which are often obtained by recording the required points of the task with the aid of a teach box (see Chap. 7). The resulting position commands are sent to the corresponding control loops and the motion starts. When the required posi-

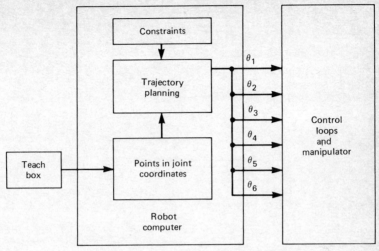

FIG. 2-4   A schematic diagram of PTP articulated robot system.

tions are reached by all axes of motion, the next set of joint coordinates is read and the operation proceeds.

The term *trajectory planning* means the determination of the actual trajectory, or path, along which the robot end effector will move. In trajectory planning of PTP robots two types of constraints are taken into account:

1. The maximum admissible acceleration and velocity at each joint
2. Constraints arising from arm geometry and work space limitations.

To demonstrate trajectory planning, let us take a simple case in which each joint of a PTP robot must move to its end point θ_f as fast as possible, without exceeding a maximum admissible acceleration a_m and a maximum velocity V_m. The desired trajectory is shown in Fig. 2-5, and it consists of three regions: an initial segment $[0, t_1]$ which is based on maximum acceleration, an intermediate region $[t_1, t_2]$ in which the joint moves in its maximum admissible velocity, and a final segment $[t_2, t_f]$ for deceleration. For simplicity we assume that the time durations of the acceleration and deceleration segments are equal

$$t_1 = t_f - t_2 \tag{2-4}$$

The control program must calculate the two switching times t_1 and t_2 based upon the initial and final angular values ($\theta_0 = 0$ and θ_f) of each joint and the maximum admissible velocity V_m and acceleration a_m.

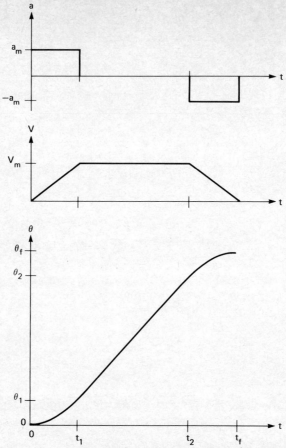

FIG. 2-5   Typical acceleration, velocity and position of an axial motion in PTP robot.

The trajectory during the first segment is

$$\theta(t) = \frac{a_m t^2}{2}$$

(2-5)

The corresponding final point is

$$\theta_1 = \frac{a_m t_1^2}{2}$$

(2-6)

and the velocity is

$$V_m = a_m t_1$$

(2-7)

The trajectory during the second segment is

$$\theta(t) = \theta_1 + V_m(t - t_1) \tag{2-8}$$

and its final point is

$$\theta_2 = \theta_1 + V_m(t_2 - t_1) \tag{2-9}$$

The joint position in the third region is

$$\theta(t) = \theta_2 + V_m(t - t_2) - \frac{a_m(t - t_2)^2}{2} \tag{2-10}$$

The end point is calculated by substituting $t = t_f$ in Eq. (2-10) and using Eq. (2-4), which yields

$$\theta_f = \theta_2 + V_m t_1 - \frac{a_m t_1^2}{2} \tag{2-11}$$

Combining Eqs. (2-6), (2-7), (2-9), and (2-11) gives

$$\theta_f = V_m t_2 \tag{2-12}$$

The two required switching times are calculated from Eqs. (2-7) and (2-12):

$$t_1 = \frac{V_m}{a_m} \tag{2-13}$$

and

$$t_2 = \frac{\theta_f}{V_m} \tag{2-14}$$

In the interval $[0, t_1]$ the axis accelerates, then moves at constant velocity, and in the interval $[t_2, t_f]$ it decelerates. The joint motion is terminated at time t_f, where

$$t_f = \frac{V_m}{a_m} + \frac{\theta_f}{V_m} \tag{2-15}$$

In PTP robots which operate with a simultaneous motion termination in all axes, trajectory planning is more complicated. In this case only one axis performs with the switching strategy given by Eqs. (2-13) and (2-14), and the other axes are synchronized to this leading axis. The leading axis is the one with the slowest t_f according to Eq. (2-15) and is the only one which moves at the maximum velocity V_m. The velocities of the other axes are calculated according to Eq. (2-15) with a given t_f. PTP motions in which the joints complete their motion simultaneously are sometimes denoted as joint-interpolated motions (Unimation's PUMA), PTP-linear (ASEA, Inc.), or coordinated axis control.

The commands to each axis of motion in PTP robots consist of a velocity signal and the joint end-point coordinate. Typically, the velocity signal is sent to each loop according to the following strategy:

$$
V(t) = \begin{cases}
a_m t & 0 \leq t \leq t_1 \\[4pt]
V_m & t_1 \leq t \leq t_2 \\[4pt]
V_m - a_m t & t_2 \leq t \leq t_f
\end{cases}
\tag{2-16}
$$

The desired end-point coordinate θ_f is sent to a downcounter. The counter is decremented by pulses from the encoder and continuously shows the value $\theta_f - \theta(t)$. When the counter reaches a zero, the required position is obtained.

The velocity switching points in Eq. (2-16) can be calculated on the basis of velocity and distance rather than time. From Eq. (2-13), a switching time occurs when the velocity command $a_m t$ reaches the value V_m, which is used for determining the first switching point. The second switching point is calculated by substituting Eq. (2-13) into (2-11), which yields

$$
\theta_f - \theta_2 = \frac{V_m^2}{2a_m}
\tag{2-17}
$$

Consequently, when the position downcounter reaches the value given in Eq. (2-17) the velocity command is changed from $V = V_m$, becoming $V = V_m - a_m t$.

With CP robots the calculation of the command signals to the control loops is more complicated, since the arm must move toward the end point along a specified trajectory. The end points in CP robots are presented in world coordinates, namely the x, y, and z values as related to the robot base are given in the robot computer. In cartesian-type robots these values are also the position commands to the control loops, but in noncartesian robots, a coordinate transformation and interpolation are required.

When interpolating, the computer first divides (in real time) the trajectory between the end points into small sections along the same straight line, as shown in Fig. 2-6. It is worthwhile to note that desired trajectories in robots are usually straight lines. The motion from the beginning to the end of each section is obtained by solving the inverse kinematics problem for the section's end point, namely transforming the point coordinates to corresponding joint commands. For the trajec-

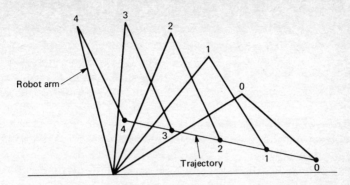

FIG. 2-6 Reference points are added along a trajectory of a CP robot.

tory in Fig. 2-6, the robot computer sends reference commands to the axes of motion at points 1, 2, 3, and 4, and the robot moves successively from point to point. The path between successive points is not predicted and is usually not a straight line.

The need to divide the path into small sections arises from the inability of noncartesian robot interpolators to provide linear motion between the end points. By interpolating along section points in straight line, the effective motion becomes nearly linear. The spacing between points, the linearity obtained, and the velocity at which the robot moves all depend very much on the computational speed of the interpolation method, as discussed in detail in Chap. 5. Robot manufacturers zealously guard their interpolation algorithms.

The location of the extra points along the trajectory is based upon a motion during equal time intervals which are determined by the system designer. A typical value can be $T_a = 30$ ms, and in this case, new reference commands are sent to the control loops every 30 ms. Therefore, the distance Δl between two successive points along the trajectory depends on T_a and the required velocity V

$$\Delta l = V T_a \qquad (2\text{-}18)$$

The components of Δl for a straight line trajectory are

$$\Delta x = V_x T_a$$

$$\Delta y = V_y T_a \qquad (2\text{-}19)$$

$$\Delta z = V_z T_a$$

where V_x, V_y, and V_z are given in Eqs. (2-1) through (2-3).

2.2 CONTROL LOOPS OF ROBOTIC SYSTEMS

Control systems can operate either in an open loop or in a closed loop. In open-loop control systems the output has no effect upon the input. As an example of an open-loop system, assume that a constant voltage is applied to an electric motor and consequently the motor rotates. The speed of the motor's shaft is the output, and the supplied voltage is the input. A load on the motor will cause a speed decrease, a situation which cannot be remedied since the input voltage is not affected by the speed variations.

A better system will be one in which the output is sensed and fed back to be compared with the input variable. In our example, the motor speed can be sensed and converted to voltage with the aid of a tachometer, and then this voltage can be compared with the input voltage. Based on this comparison, any necessary corrective action automatically takes place in order to return the output speed to the desired value. Systems in which the output affects the input to the controlled element are called *closed-loop control systems.*

Each axis of motion of the robot arm is separately actuated by a control loop which contains a drive element. In closed-loop systems, the resultant motion is sensed by a feedback device, such as an incremental encoder. The axial drive may be a dc servomotor, a stepping motor, a hydraulic actuator, or a pneumatic cylinder. The type selected is determined mainly by the accuracy and power requirements of the robot. The drive elements may be coupled directly to the robot joints or may drive them indirectly through gears, chains, cables, or leadscrews.

Open-loop robots use stepping motors for driving the axes, as shown in Fig. 2-7. A stepping motor is a device whose output shaft rotates through a fixed angle in response to an input pulse. Stepping motors provide the simplest means of converting electrical pulses into proportional angular movement, and as a result represent a relatively inex-

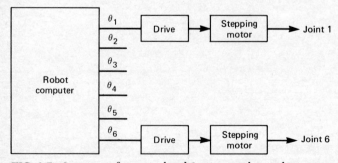

FIG. 2-7 Structure of an articulated 6-axis open-loop robot system.

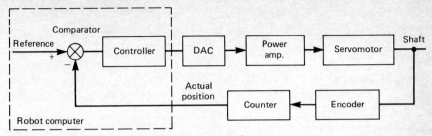

FIG. 2-8   Block diagram of a closed-loop control system.

pensive solution to the control problem. Since there is no feedback from the shaft position, positioning accuracy is solely a function of the motor's ability to step through the exact number of pulses provided at its input.

One of the main characteristics of a stepping motor is that its maximum velocity depends upon the load torque. The higher the load torque, the smaller the maximum allowable velocity of the motor. As a consequence, stepping motors cannot be used in systems with variable-load torques, since an unpredictably large load causes the motor to loose steps resulting in unintended position error. In robotic systems, however, motors must develop torques that depend not only on the gripper load but also on the position of the robot arm. Therefore, stepping motors are not recommended for use as drives for industrial robots.

Besides a few laboratory robots which use stepping motors in open-loop control, most modern robots function in closed-loop control, where the actual position of each axis of motion is measured by a feedback device.

Figure 2-8 shows a closed-loop control for a single axis of motion. The motor shaft can either drive a revolute joint or be coupled to a leadscrew to drive a linear axis. The closed-loop control measures the actual position and velocity of each axis and compares them with the desired values. The difference between the actual and the desired values is the error. The control strategy is designed to eliminate this error or reduce it to a minimum, as in the case of closed-loop negative feedback systems.

The feedback device, which is an incremental encoder in Fig. 2-8, is mounted on the shaft and supplies a pulsating output. The comparator correlates the input and the feedback and gives, by means of a digital-to-analog converter (DAC), a signal representing the position error of the axis. The error signal is used to drive the dc servomotor.

The incremental encoder is the most popular feedback device in robot systems. An encoder is attached to each joint of the robot manip-

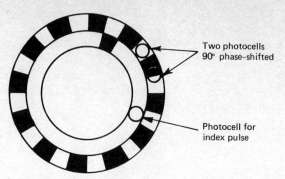

Two photocells
90° phase-shifted

Photocell for
index pulse

FIG. 2-9   Incremental encoder.

ulator. The encoder is a shaft-driven device delivering pulses at its output terminals. It contains a rotating disk divided into segments, which are alternately opaque and transparent as shown in Fig. 2-9. A light source is placed on one side of the disk and a photocell on the other side. When the disk rotates, each change in light intensity falling on the photocell produces an output pulse. The rate of these pulses per minute is proportional to the joint shaft speed in revolutions per minute, and the number of these pulses is proportional to the displacement of the joint.

The direction of rotation may be sensed by using an encoder with two photocells reading the same disk. The photocells are arranged so that their outputs have a 90° shift to each other, as shown in Fig. 2-9. The direction of rotation can be determined by external logic circuitry, fed by these two sequences of pulses. An additional index pulse can be made available by a separate zone containing only a single clear section provided on the disk. The index pulse serves as a zero reference position when the robot system is switched on.

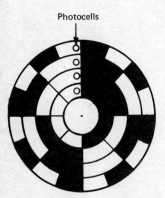

Photocells

FIG. 2-10   Absolute encoder with binary code.

One disadvantage of using incremental encoders to determine position is the possibility of incorrect data resulting from false counts being generated by electrical noise, transients, or other outside disturbances. Gross errors can also result from power interruption. Those errors are eliminated by using absolute encoders. Absolute rotary encoders use a multiple-track disk which defines the shaft position by means of a binary word (Fig. 2-10) or another code, such as the Gray code. The reading system employs a lamp and photocells to detect the light which passes through the transparent portions of the disk. A photocell is provided for each track on the disk. The output from all cells gives the actual shaft position of the joint.

In the closed-loop system of Fig. 2-8 each control loop is closed through the robot computer itself. The encoder pulses are accumulated in a counter, which is sampled by the computer at constant time intervals. In other system configurations, a microprocessor is used as the comparator in each control loop, and a minicomputer coordinates the reference commands to the individual microprocessors.

Extreme care must be taken during the design of a closed-loop control system. By increasing the magnitude of the feedback signal (more pulses per revolution of the encoder shaft) the loop is made more sensitive. That is known as increasing the open-loop gain. Increasing the open-loop gain excessively may cause the closed-loop system to become unstable, which obviously should be avoided.

The design of the control loop and the choice of the drive element require a knowledge of the nature of the application and the loading torques. The allowable positioning error, accuracy, repeatability, and response time also have to be taken into consideration when an optimum performance is required.

2.3 THE MANIPULATOR

The manipulator is the mechanical unit which performs the movement function in the robot. It consists of a series of mechanical links and joints capable of producing controlled movement in various directions. The manipulator is composed of the main frame (the arm) and the wrist, each having three degrees of freedom, or axes of motion. There are applications, however, which require less than six degrees of freedom, and in these cases the number of wrist motions is reduced.

Structurally, the robots can be classified according to the coordinate system of the main frame:

Cartesian: three linear axes

Cylindrical: two linear and one rotary axis

Spherical: one linear and two rotary axes

Articulated or Jointed: three rotary axes

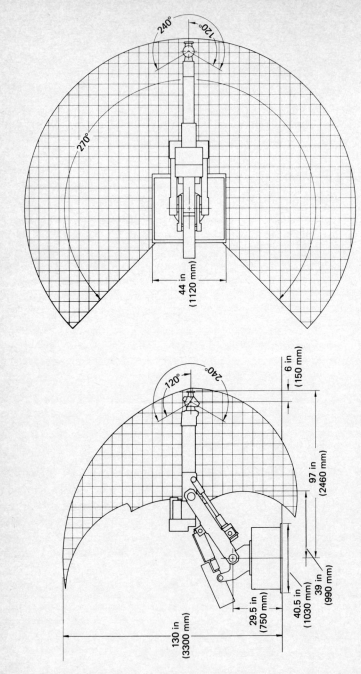

FIG. 2-11 Reach envelope of industrial robot. (Cincinnati Milacron.)

Different applications may require different coordinate systems. For example, a cylindrical robot might be best suited to a straight load of a punch press, while an articulated arm might be best for reaching a weld seam which is located behind a structure and is invisible from the robot base.

In order to aid in defining the coordinate system, a symbolic notation is often used to describe the type and number of each joint, starting from the base to the end of the arm. Linear joints can be sliding or prismatic, designated S or P, and revolute joints, designated R. By this notation the spherical robotic arm, for example, would be called an RRP arm, and the articulated one, an RRR arm.

One of the most important performance characteristics of a manipulator is the shape of its *reach envelope,* or its *work volume,* which is shown in Fig. 2-11. The shape of the work volume depends on the coordinate system, and its size depends on the dimensions of the robot arm. It should be noted that when a gripper or another tool is attached to the wrist, the work volume exceeds the one given by the robot manufacturer, and this should be taken into account when planning for the safety of the people working near the robot.

The various configurations of robotic arms and their corresponding work volumes are discussed below.

2.3.1 Cartesian Coordinate Robots

The main frame of cartesian coordinate robots consists of three orthogonal linear (prismatic or sliding) axes, as shown in Fig. 2-12. The structure can be more rigid if constructed like the work table of a milling machine, but then the ratio between the robot work volume and floor space becomes smaller. In a cartesian robot, the manipulator hardware and the control program are similar to those of CNC machine tools. Therefore, the arm resolution and repeatability might also be on the order of magnitude of machine tool resolution. The base axis in Fig. 2-12 has a vertical motion, but in many commercial robots this axis moves horizontally, as shown in Fig. 2-13.

The wrist of a cartesian robot

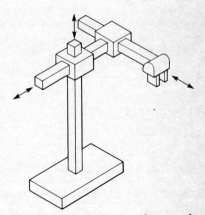

FIG. 2-12 Cartesian coordinate robot manipulator. (U.S. Air Force.)

FIG. 2-13 Cartesian coordinate robot. (Fraser Automation.)

can be programmed to trace a linear path in the space simply by driving each of the participating axes at constant velocity. If the robot's end point has to trace a straight path of l length units in a velocity V, the axial velocities are

$$V_x = (x/l)V \qquad (2\text{-}20)$$

$$V_y = (y/l)V \qquad (2\text{-}21)$$

and $$V_z = (z/l)V \qquad (2\text{-}22)$$

where $$l = \sqrt{x^2 + y^2 + z^2}$$

and x, y, and z are the components of l in the X, Y, and Z directions, respectively. Such simple relationships are not valid for other robot coordinate systems, for which complex computer algorithms are required to drive the end point along straight-line trajectories. The controlling computers of these robots usually must transform the cartesian coordinates, in which most robotic task programs are written, to the coordinates associated with the joints of the manipulator.

Another important feature of cartesian robots is equal and constant spatial resolution, that is, the resolution is fixed in all axes of motion and throughout the work volume of the robot arm. This is not the case with other coordinate systems, as will be shown later.

Despite all these attractive features, the cartesian robot is not popular in industry. The reason is that it lacks mechanical flexibility; it cannot reach objects on the floor or reach points invisible from its base. Also the speed of operation in the horizontal plane is usually slower than that associated with robots having a rotary base.

2.3.2 Cylindrical Coordinate Robots

The main frame of cylindrical coordinate robots consists of a horizontal arm mounted on a vertical column which, in turn, is mounted on a rotary base, as shown in Fig. 2-14. The horizontal arm moves in and out, the carriage moves up and down on the column, and these two units rotate as a unit on the base. Thus the working volume is a cylindrical annular space as shown in Fig. 2-15. A commercial robot with a cylindrical coordinate motion driven with hydraulic systems is shown in Fig. 2-16.

The resolution of the cylindrical robot is not constant and depends on the distance r between the column and the gripper along the horizontal arm. If the resolution unit of the rotary base is α radians, then the resolution at the arm end is αr, as shown in Fig. 2-17. This is illustrated by the following example.

EXAMPLE 2-1

The position-measuring device of the rotary axis of a cylindrical robot is an incremental encoder which emits 6000 pulses per revolution and is mounted

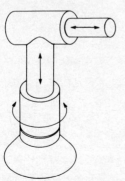

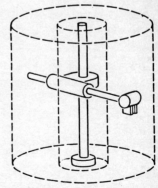

FIG. 2-14 Cylindrical coordinate manipulator.

FIG. 2-15 Work volume shape of cylindrical manipulator. (U.S. Air Force.)

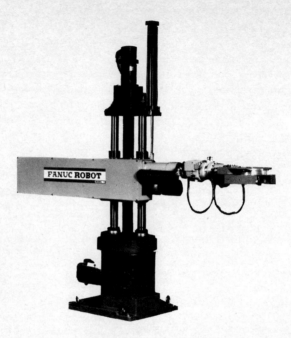

FIG. 2-16   Cylindrical coordinate robot. (Fanuc, Japan.)

directly on the shaft. If the maximum length of the horizontal arm is 1 m, what is the worst-case resolution at the arm end?

Solution.   The resolution at the base is $\alpha = 360°/6000 = 0.06°$. The resolution at the arm end is $1000 \times 0.06 \times \pi/180 = 1.05$ mm.

The result of Example 2-1 demonstrates that the arm resolution around the base might be two orders of magnitude larger than that obtained with cartesian robots or machine tools (0.01 mm). This is one of the drawbacks of cylindrical robots as compared to cartesian robots. Robots with cylindrical geometry do offer the advantage of higher speed at the end of the arm due to the rotary axis. However, this speed is limited in many robots because of the varying moment of inertia of the robot arm, which depends on the load at the robot's end effector and the position of the arm itself.

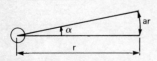

FIG. 2-17   The resolution at the arm end in robots with rotary base.

Good dynamic performance is usually difficult to achieve in robots which contain a rotary base. The torque which the base motor has to supply depends on the position, speed, and acceleration of the other joints, and this causes variations in the reflected torque and moment of inertia. The moment of inertia reflected at the base depends not only on the weight of the object being carried but also upon the distance between the base axis and the manipulated object. This distance is a function of the instantaneous position of the gripper and the other joints during the motion. As a result, the effective moment of inertia at the base drive generally varies with time or position, which consequently results in inferior dynamic performance of the arm. The effect is further discussed in Chap. 3 and is regarded as one of the main drawbacks of robots containing revolute joints.

2.3.3 Spherical Coordinate Robots

The kinematic configuration of spherical, or polar, coordinate robot arms is similar to the turret of a tank. It consists of a rotary base, an elevated pivot, and a telescoping arm which moves in and out as shown in Fig. 2-18. The magnitude of rotation is usually measured by incremental encoders mounted on the rotary axes. The work envelope is a thick spherical shell as shown in Fig. 2-19. An experimental robot for research purposes which uses a spherical coordinate system was designed by the author and built by students at the University of Michigan. It is shown in Fig. 2-20.

The disadvantage of spherical robots compared with their cartesian counterparts, is that there are two axes with relatively low resolution that varies with the arm length. This is demonstrated in the following example.

EXAMPLE 2-2

Find the worst spatial resolution of a spherical robot with 500-mm arm length. The robot is equipped with three encoders emitting 1000 pulses per revolution. The linear axis is actuated with the aid of a 10-mm pitch leadscrew, and its encoder is mounted on the leadscrew. The other two encoders are mounted through a gear ratio of 1:22.

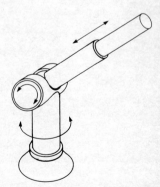

FIG. 2-18 Spherical coordinate manipulator.

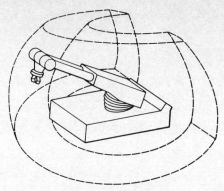

FIG. 2-19 Work volume of spherical manipulator (U.S. Air Force.)

FIG. 2-20 A spherical coordinate robot built by the author's students at the University of Michigan.

Solution.   The linear axis resolution is $10/1000 = 0.01$ mm. The rotary axes resolution is

$$(1/22) \times (360/1000) \times 500 \times \pi/180 = 0.14 \text{ mm}$$

The spatial resolution is 0.14; 0.14; 0.01 mm.

The Example 2-2 demonstrates the large difference in the obtained resolution between linear and rotary axes. Motions along rotary axes, however, are faster than those along linear axes. The main advantage of spherical robots over the cartesian and cylindrical ones is a better mechanical flexibility: the pivot axis in the vertical plane permits access to points at base level or below it.

2.3.4 Articulated Robots

Articulated robots consist of three rigid members connected by two revolute joints and mounted on a rotary base as shown in Fig. 2-21. This kinematic arrangement closely resembles that of a human arm. The gripper is analogous to the hand, which attaches to the forearm via the "wrist" (see Fig. 1-1). The "elbow" joint connects the forearm and

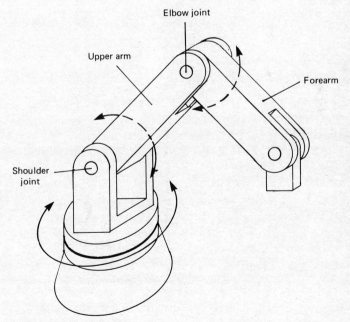

FIG. 2-21   Articulated manipulator.

FIG. 2-22   Articulated robot (Sharnoa Electronics.)

the upper arm, and the "shoulder" joint connects the upper arm to the base. Sometimes a rotary motion in the horizontal plane is also provided at the shoulder joint, as shown in Fig. 2-21. A commercial articulated robot is shown in Fig. 2-22.

Since the articulated robot has three rotary axes, its spatial resolution depends entirely on the arm position. The accuracy of an articulated robot is poor since the joint errors are accumulated at the end of the arm. On the other hand, it can move at high speeds and has excellent mechanical flexibility, which make it the most common small- and medium-sized robot.

2.3.5 Direct and Indirect Drives

The manipulator joints can be driven directly or indirectly. With direct drive, the joint shaft is coupled to the rotor of the drive motor. With indirect drive, the joint is connected to the drive motor through a transmission mechanism.

Direct drive might provide better positioning accuracy since the intermediate gearing is eliminated and consequently the mechanism is free of backlash and hysteresis. Another advantage is the improved reliability because of the smaller number of mechanical parts. However, the main drawback of direct-drive manipulators is that the motors which drive the joints are themselves loads for the motors at the lower joints (i.e., joints closer to the base). For example, the three motors at the wrist are loads for the motor at the elbow joint, and all these four motors are loads for the shoulder joint, etc. As a result, the payload is very small compared with the arm weight, and therefore most commercial robots use indirect-drive mechanisms.

A five-axis robot manipulator with indirect drive which was designed and constructed in the Robotics Laboratory of the Technion is shown in Fig. 2-23. The arm of this robot has three degrees of freedom which provide spherical motion. It consists of a rotary base, an elevated pivot driven with the aid of a leadscrew connected to its end, and an articulated forearm driven through a four-bar mechanism actuated by another leadscrew. The two motors of the leadscrews as well as the two motors of the wrist are mounted on the rotary base.

The articulated structure of the forearm permits a remote control of the two wrist motions, an approach which is difficult to implement in regular spherical robots. With this structure, the mechanical advantages of an articulated arm are combined with those of a spherical one and produce a high-performance robot.

The removal of actuators from the arm and the wrist reduces the weight of the upper part of the manipulator. In addition, a structure which contains an indirect-drive mechanism has higher rigidity. These two features enable good dynamic performance and good positioning accuracy. Moreover, since the weight of actuators becomes of minor significance, they can be more powerful, resulting in faster motions and higher payload capacity of the robot.

In addition to its lighter weight, the advantage of the remote-driven wrist is that its orientation is defined relative to the rotary base, independently of the arm's orientation.

As was mentioned, two of the

FIG. 2-23 Spherical robot with articulated forearm employing indirect drives. The robot was built by the author's students at the Technion.

axial actuators of the manipulator of Fig. 2-23 are leadscrews. Comparing them to other gearing systems, such as worm gear or harmonic drive, the leadscrew mechanism provides a zero backlash (using a ball-screw with double nut configuration) and a stiffer driving system. If a nonreversible leadscrew is used (i.e., a self-locked leadscrew), a joint brake is not needed, and there are no motor overheating problems at zero speeds.

2.4 THE WRIST MOTIONS AND THE GRIPPER

The end effector is connected to the main frame of the robot through the wrist. A typical wrist including three rotary axes allowing roll, pitch, and yaw was shown in Fig. 1-1. In this wrist the roll is a rotation in a plane perpendicular to the end of the arm, pitch (or bend) is a rotation in a vertical plane, and yaw is a rotation in a horizontal plane through the arm. Another popular wrist allowing a sequence of roll, bend, and again roll is shown in Fig. 8-7 (the PUMA's wrist). Although most wrists use three rotary axes, there are applications which require only two axes of motion in the wrist. For example, since a welding gun is a symmetrical tool, most arc-welding tasks require a wrist with only two degrees of freedom.

The wrist should be designed to be as light as possible. Reductions of weight at the wrist increases the maximum allowable load and reduces the moment of inertia, which improves the dynamic performance of the robot arm. In order to reduce weight at the wrist, the wrist drives are sometimes located at the base and the motion is transferred with chains or rigid links as shown in Fig. 2-24.

The design requirements of the wrist depend on the application. A wrist for process applications, such as arc welding and spray painting,

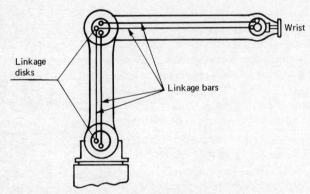

FIG. 2-24 Schematic diagram of wrist transmission method.

is designed differently from a wrist for material handling. Many process application tasks are performed in confined areas, such as weldments in autobodies, and in order to perform them efficiently a compact wrist with a high level of dexterity and large angular range of orientation is required. It is also typical that the payload for most process applications is constant and small relative to that required in machine tool loading and material handling.

End effectors fall into two categories: grippers and tools for process applications, such as welding torches, painting guns, drills, and grinders. Grippers are used in handling, machine loading, and assembly applications.

In most grippers the mechanism is actuated by a pneumatic piston which moves the gripper fingers. When the robot is handling glass products or parts with highly polished surfaces, a vacuum-type gripper can be used. Electromagnetic grippers have not been widely used, as they can present problems by picking up chips when loading machine tools, or picking up more than one part in assembly and handling.

In many cases, whole families of parts can be manipulated by one gripper with interchangeable fingers, which are attached to the gripper body. In these cases, care should be taken in designing the gripper with long-life fasteners for the fingers to ensure good repeatability. The repeatability of the gripper is also determined by its mounting to the wrist. Most wrists are equipped with a flange to which a gripper or another end effector can be mounted.

Repeatability depends also on the contact area between the gripper and the part. The friction coefficient between the two surfaces and the clamping force should be high enough to avoid sliding. To guarantee better clamping the gripper should be designed according to the shape of the part. For example, when gripping cylindrical parts from the outside, the fingers should have a V shape to guarantee better contact with the part.

Care should be given to the design of the gripping pads of the fingers. A material which is widely used for gripping pads is a polyurethane bonded on steel (Mutter, 1983). The polyurethane can be shaped or machined to any configuration and will stand up to thousands of hours of compression and release. It also has a very high coefficient of friction and will not mark the part.

A consideration should also be given to the gripping surfaces on the part. These surfaces should be allowed minor scratches or marks, and must be suitable for gripping. The gripped surfaces must be nearly parallel for a two-finger gripper, otherwise a gripper with three fingers should be used. Note that the opening of the gripper at any wrist position must be greater than the width of the gripped part.

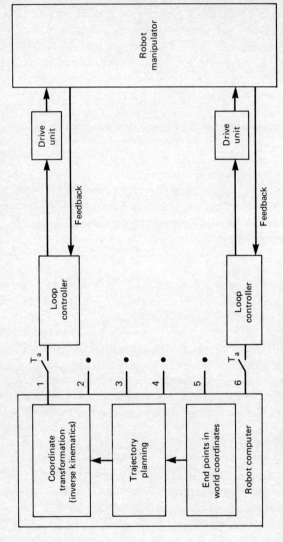

FIG. 2-25 Conceptual structure of CP robot system.

2.5 STRUCTURE OF CONTINUOUS-PATH ROBOT SYSTEMS

CP robots can operate in either PTP or CP mode. The CP mode is used to obtain a geometrically defined trajectory in space, but in practice only straight-line motions can be performed by CP robots. (By contrast, CNC machine tools can move along straight lines and circular arcs.) A straight-line motion is useful when the robot tool must move relative to an external object. An example is a robot performing a precision insertion of a pin in assembly or loading a workpiece into the chucks of a lathe.

The conceptual structure of a six-axis CP robot system is illustrated in Fig. 2-25. The task of the robot manipulator is defined by a sequence of end points defining the termination of each trajectory. These end points are stored in the task program section of the robot computer. In CP operations the end points and the tool orientation are given in world coordinates. That means that the x, y, and z values of the TCP and the orientation angles of the robot tool are given in a reference frame fixed in the robot base and denoted as the world coordinate system.

The robot computer performs trajectory planning and transformation from world coordinates to joint coordinates, as shown in Fig. 2-25. The trajectory planning algorithm inserts intermediate tool locations in a straight line (see Fig. 2-6) and determines the acceleration and deceleration periods in each trajectory. The coordinate transformation routine rapidly transforms the intermediate tool locations to joint commands by applying inverse kinematic algorithms. The resultant motion becomes nearly linear.

In CNC terminology, the term *interpolator* is used for a computer algorithm which inserts intermediate points between the segment's end points and generates reference commands to the control loops in order to coordinate their motion to obtain a required path (Koren, 1983). For example, the circular interpolator of a CNC milling machine receives the end point and the radius of an arc and emits reference commands proportional to $\cos \omega t$ and $\sin \omega t$, which are sent to the control loops of the X and Y axes, respectively. The equivalent assignment in robotics is performed by the combination of the trajectory planning and coordinate transformation routines, and therefore their combined action is denoted in this text as *interpolation.*

The joint positions and velocities produced by the interpolator are sent as references to the control loops at time intervals T_a. This time interval is dictated by the time required for one iteration of the interpolation. Each axis of motion of the robot is controlled by a separate loop. In modern robots, a microprocessor is included as an integral part of each loop controller. The controller compares the reference and the feedback signals and determines the instantaneous command to the drive unit. The drive unit contains a power amplifier, an actuator (e.g., a dc servomotor), and often also a tachometer as an auxilliary feedback device, which closes an internal loop in the drive unit. The main axial feedback device, however, is either an encoder or a resolver which senses the actual axial position and transmits it to the loop controller in order to close the main control loop. The detailed structure of control loops is given in Chap. 3. Coordinate transformation and interpolation techniques are presented in Chaps. 4 and 5.

REFERENCES

Ardayfio, D. D., and H. J. Pottinger: "Computer Control of Robotic Manipulators," *Mech. Eng.*, vol. 107, August 1982, pp. 40–45.

Duffy, J.: *Analysis of Mechanisms and Robot Manipulators*, John Wiley & Sons, New York, 1980, chap. 1.

Engelberber, J. F.: *Robotics in Practice*, AMA/COM, 1980, chap. 2.

Koren, Y.: *Computer Control of Manufacturing Systems*, McGraw-Hill Book Co., New York, 1983, chap. 9.

Mutter, R. F.: "Tailoring the End Effector to the Task," *Robot. Today*, vol. 5, no. 5, October 1983, pp. 69–72.

U.S. Air Force: *ICAM—Robotics Application Guide*, Report AFWAL-TR-80-4042, vol. II, April 1980.

CHAPTER 3

Drives and Control Systems

Every axis of the robot manipulator includes a drive which converts the electrical command signals of the computer to mechanical motions. In most computerized robots, the axial motions are monitored and controlled by closed-loop systems, which compare references with feedback signals to determine the axial errors. These errors are amplified and used to generate the drive motions.

Drives for computerized robot systems are usually either hydraulic or electric servomotors. A few robot manufacturers employ stepping motors as drives; however, in the author's opinion, stepping motors are not appropriate drives for robot arms. The allowable speed of a stepping motor is a function of its load torque, but the latter strongly depends on the arm position and the gripper load. An excessive load on the stepping motor might cause a subsequent loss of steps. In addition, stepping motors are limited in resolution and tend to be noisy. For all of these reasons, stepping motors are seldom used in robots and will not be further discussed in this text.

DC servomotors provide excellent speed regulation, high torque, and high efficiency, and therefore they are ideally suited for control applications. DC motors can be designed to meet a wide range of power requirements and are utilized in most small- to medium-size robots. Hydraulic systems are well suited for large robots, where power requirements are high. The cost of hydraulic drive is not proportional to the power required, and thus they are expensive for small- to medium-size robots.

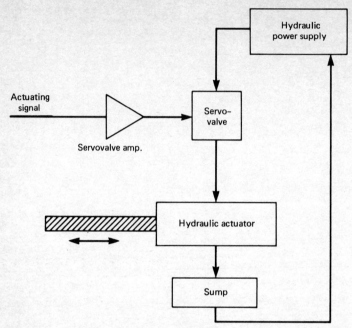

FIG. 3-1   Hydraulic system for a robot arm.

3.1 HYDRAULIC SYSTEMS

Hydraulic systems are used extensively for driving high-power robots, since they can deliver large power while being relatively small in size. They can develop much higher maximum angular acceleration than dc motors on the same peak power. They have small time constants, and this results in smooth operation of the robot axes.

Hydraulic systems, however, present some problems in terms of maintenance and leakage of oil from the transmission lines and the system components. The oil must be kept clean and protected against contamination. Other undesirable features are the dynamic lags caused by the transmission lines and viscosity variations with oil temperature.

As shown in Fig. 3-1, hydraulic systems generally comprise the following components: (1) a hydraulic power supply, (2) a servovalve for each axis of motion, (3) a sump, and (4) a hydraulic motor for each axis of motion.

3.1.1 Hydraulic Power Supply

The hydraulic power supply is a source of high-pressure oil for the hydraulic motor and the servovalve. The main components of the hydraulic power supply are shown in Fig. 3-2 and are as follows:

1. A pump for supplying the high-pressure oil. The frequently used types are the gear pump and radial or axial displacement pumps.
2. An electric motor, usually a three-phase induction motor, for driving the pump.
3. A fine filter for protecting the servosystem from any dirt or chips.
4. A coarse filter, located at the input of the pump, for protecting the latter against contamination that has entered into the oil supply.
5. A check valve for eliminating a reverse flow from the accumulator into the pump.
6. A pressure-regulating valve for controlling the supply pressure to the servosystem.
7. An accumulator for storing hydraulic energy and for smoothing the pulsating flow. Accumulators can provide a large amount of energy over a short interval of time and are used where the load is characterized by an average demand which is far below the required peak. The accumulator supplies the peak requirements and is subsequently recharged by the pump. Another function of the accumulator is to smooth the pulsations caused by the pump, and the variations caused by the sudden motions of the valve. The accumulator functions like a capacitor in an electric circuit.

3.1.2 Servovalve

The electrohydraulic servovalve controls the flow of the high-pressure oil to the hydraulic motor. The servovalve receives a voltage-actuating signal and uses it to drive either an electric motor or a solenoid device,

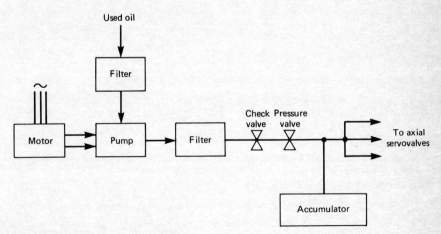

FIG. 3-2 Hydraulic power supply.

which moves the valve spool. The magnitude of the input voltage (or current) defines the flow rate of oil through the valve. The flow rate of oil through the valve is proportional to the velocity of the hydraulic motor. The time constant of a servovalve, in a high-power system, is on the order of 5 ms, and is usually negligible compared with the other lags in the system.

3.1.3 The Sump

The used oil is returned to a sump, or tank, through a special return line. The oil is fed back to the hydraulic power supply and forms its source of fluid.

3.1.4 The Hydraulic Motor

The hydraulic motor is either a hydraulic cylinder for linear motion or a rotary-type motor for angular motion.

The hydraulic cylinder, due to the large quantity of high-pressure oil which it contains, is limited to a relatively small motion. The rotary hydraulic motor is usually used in larger power servosystems. It operates at high speeds and is geared down to the shaft which drives the robot joint.

3.2 DIRECT-CURRENT SERVOMOTORS

DC motors allow precise control of either the speed, by manipulation of the voltage, or the torque, by manipulation of the current applied to the motor. They are ideally suited for driving the axes of small- to medium-size robots.

3.2.1 Principle of Operation

The dc motor is actually a dc machine, which can function either as a motor or as a generator. The principle of operation of a dc machine is based on the rotation of an armature winding within a magnetic field. The armature is the rotating member, or *rotor*, and the field winding is the stationary member, or *stator*. The armature winding is connected to a commutator, which is a cylinder of insulated copper segments mounted on the rotor shaft. Stationary carbon brushes which are connected to the machine terminals are held against the commutator surface and enable the transfer of direct current to the rotating winding. For the case in which the dc machine serves as a motor, electrical energy is supplied to the armature from an external dc source, and the motor converts it to mechanical energy.

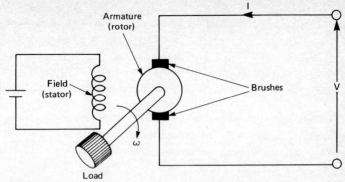

FIG. 3-3 Schematic diagram of a separately excited dc motor.

The dc motors which are used in robots are of a servomotor type. The field flux in dc servomotors is constant. This is achieved by either connecting the field permanently to a constant dc source (see Fig. 3-3) or using a permanent magnet for the motor's field. Two equations are required to define the behavior of a dc servomotor: the torque and the voltage equation. The torque equation relates the torque to the armature current

$$T = K_t I \qquad (3\text{-}1)$$

and the voltage equation relates the induced voltage in the armature winding to the rotational speed

$$E = K_v \omega \qquad (3\text{-}2)$$

where T = magnetic torque, N·m (in SI units)
I = current in armature circuit, A
E = induced voltage, V
ω = angular velocity, rad/s

The parameters K_t and K_v are referred to as the *torque* and *voltage constants*. In SI units the torque constant in newton-meters per ampere equals the voltage constant in volt-seconds per radian. If U.S. customary units are applied, the values of K_t and K_v become different. Many practical designs used technical mks units, in which the speed is given in revolutions per minute and the torque in kilogram-meters. As a consequence, K_t is given in kilogram-meters per ampere, and K_v in volts per revolution per minute. The ratio between these two constants is

$$K_v/K_t = 2\pi g/60 = 1.025$$

and in practice they are treated as equal constants.

For a motor, an input voltage V is supplied to the armature, and the corresponding voltage equation becomes

$$V - IR = K_v\omega \tag{3-3}$$

where R = resistance of armature circuit and IR = voltage drop across this resistance. The armature inductance is negligible in Eq. (3-3).

Multiplying Eqs. (3-1) and (3-3) yields the power equation

$$P = \omega T = VI - I^2R$$

where P = mechanical output power
 VI = electric input power
 I^2R = electric power loss

3.2.2 Dynamic Response

The dc servomotors which drive the manipulator axes are loaded by torques consisting of dynamic and static components. Dynamic torques are caused by the motion of the robot arm. There are three types of dynamic torques. (1) Inertial torques which are proportional to the acceleration. They are caused by the acceleration of the driven joint itself and by the accelerations of the other robot joints. (2) Coriolis torques which are proportional to the product of two joint velocities. (3) Centripetal torques which are proportional to the square of other joint velocities and are caused by a rotation of a link around a point. Static torques in robotics are mainly caused by the gravity force and are proportional to the vertical gravity component. Static torques are also caused by friction in the gears, leadscrews, and other transfer mechanisms. In assembly and machining tasks an external torque might act on the end effector and consequently generate additional load torques on the motors.

As an example consider the two-link planar manipulator in Fig. 3-4. Assume that joint 1 is directly driven by a dc servomotor. The torque equation of joint 1 is given (Paul, 1981) by

$$T = J\frac{d\omega}{dt} + J_c\frac{d\omega_2}{dt} + C\omega_2\omega + D\omega_2^2 + T_g \tag{3-4}$$

where J is the combined moment of inertia of motor and manipulator and depends on the instantaneous angle between the two links (θ_2); J_c is the coupling inertia between the two joints and again depends on the angle between the links; $C\omega_2\omega$ is caused by Coriolis force; $D\omega_2^2$ is contributed by the centripetal force due to the velocity in the other joint; and T_g is the torque due to gravity.

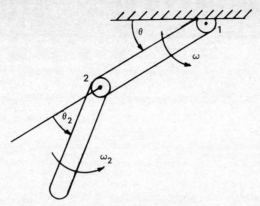

FIG. 3-4   A two-link planar manipulator.

Similar to Eq. (3-4), the torque equation at any joint of a manipulator has the following format:

$$T = J(\theta_i) \frac{d\omega}{dt} + B(\omega_i)\omega + T_s(\dot{\omega}_i, \omega_i, \theta_i) \qquad (3\text{-}5)$$

where θ_i, ω_i, and $\dot{\omega}_i$ are the angular position, velocity, and acceleration of the other robot joints, and T_s contains the coupling inertia and centripetal torque components as well as the static torques due to gravity and friction. The axial motor generates a torque, given by Eq. (3-1), to overcome the load torque given in Eq. (3-5).

Elimination of I and T from Eqs. (3-1), (3-3), and (3-5), and rearrangement of the terms so as to separate the independent variables, gives the speed equation

$$\frac{JR}{K_t} \frac{d\omega}{dt} + K_v\,\omega + \frac{RB\omega}{K_t} = V - \frac{R}{K_t}\,T_s \qquad (3\text{-}6)$$

The third term in Eq. (3-6) is negligible since

$$K_v\omega \gg IR \frac{B\omega}{IK_t}$$

$B\omega/IK_t$ represents the ratio between the Coriolis torque and the total load torque which in practice is smaller than 10 percent, and IR represents the voltage losses in the motor which are much smaller than the electromotive force $E = K_v\omega$. Note that if one prefers to take the third term into account, the following mathematical analysis is still valid; only the value of the parameter K_v is slightly increased. As a consequence, Eq. (3-6) is written as

$$\tau \frac{d\omega}{dt} + \omega = K_m V - \frac{K_m R}{K_t} T_s \qquad (3\text{-}7)$$

where K_m is the gain of the motor and is defined by $K_m = 1/K_v$, and τ is the mechanical time constant of the drive and is defined by

$$\tau = J \frac{R}{K_t K_v} \qquad (3\text{-}8)$$

The time constant depends on the moment of inertia of the robot arm, which is affected by its position in space.

The Laplace transform of Eq. (3-7) is

$$\omega(s) = \frac{K_m V(s) - [RK_m/K_t] T_s(s)}{1 + s\tau} \qquad (3\text{-}9)$$

The solution of Eq. (3-9) in the time plane depends on the applied voltage and load torque. For example, assuming that the motor is initially at rest, $T_s = 0$, and a step voltage of V is applied at the armature terminals, the solution is

$$\omega(t) = K_m V(l - e^{-t/\tau})$$

Thus the motor response is described by a steady-state speed $K_m V$, and a decaying exponential with a time constant τ given by Eq. (3-8).

EXAMPLE 3-1

The voltage and torque constants of a dc servomotor are $K_v = 0.824$ V · s/rad, and $K_t = 7.29$ lb·in/A. The armature resistance is 0.41 Ω and the armature inertia is 0.19 lb·in·s^2.
(a) Show that the voltage and torque constants are equal in SI units.
(b) Calculate the mechanical time constant of the motor.
(c) Calculate the steady-state speed for 85 V input at no-load and full-load (120 lb·in) conditions.

Solution.   (a)   The torque constant is converted to SI units as follows:

$$K_t = 7.29 \times 0.0254 \times \frac{9.81}{2.205} = 0.824 \text{ N·m/A}$$

and is equal to the voltage constant. Note that $g = 9.81$ m/s^2.
(b) From Eq. (3-8)

$$\tau = \frac{0.19 \times 0.41}{0.824 \times 7.29} = 13 \text{ ms}$$

(c) At no-load conditions

$$\omega = K_m V = \frac{85}{0.824} = 103 \text{ rad/s}$$

or

$$\omega = 103 \times \frac{60}{2\pi} = 985 \text{ rpm}$$

For a loaded motor the change in speed, from Eq. (3-6), is

$$\Delta\omega = \frac{RT_s}{K_t K_v} = -\frac{0.41 \times 120}{7.29 \times 0.824} = -8.2 \text{ rad/s or } -78 \text{ rpm}$$

The motor speed is reduced to 907 rpm.

3.2.3 Gearing

In many robots the joints are driven through a gear mechanism. The gear ratio K_g is defined as the ratio between the speed of the joint ω_j to the speed of the motor.

$$K_g = \frac{\omega_j}{\omega} \tag{3-10}$$

In order to calculate the time constant of the drive by Eq. (3-8), the inertia of the joint should be referred to the motor shaft. Consequently, the inertia J in Eq. (3-8) is

$$J = J_r + K_g^2 J_l \tag{3-11}$$

where J_r is the inertia of the rotor and J_l is the inertia of the load. Note that load torques should also be referred to the motor shaft:

$$T_s = K_g T_l \tag{3-12}$$

where T_l is the load torque at the robot joint.

3.3 CONTROL APPROACHES OF ROBOTS

Most small- to medium-size robots utilize dc servomotor actuators. Two alternative approaches exist to the control of the motion of a robot arm driven by dc motors (Koren & Ulsoy, 1982). The approach used by several United States robot manufacturers and researchers at United States universities[†] is to control the torque of the robot arm by manipulating

---

[†]Information based on personal communications of the author with Mr. V. Scheinman and Professors T. Binford (Stanford University), S. Dubowsky (MIT), G. Lee (University of Michigan), and D. Tesar (University of Florida).

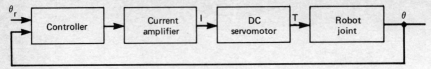

FIG. 3-5   Control loop utilizing a current amplifier.

the motor current. Another approach, commonly used by European and Japanese robot manufacturers and in NC machine tools is to control the motor rotational speed by manipulation of the motor voltage. The first approach treats the torque produced by the motor as an input to the robot joint (see Fig. 3-5). The second approach treats the robot arm as a load disturbance acting on the motor's shaft (see Fig. 3-6). This basic distinction is not merely a philosophical one and has important practical consequences for the final control system design.

A straightforward approach to the control of robot arm motion is to apply at each joint the necessary torque to move the manipulated object and to overcome friction, gravity forces, and dynamic torques due to the moment of inertia. This approach is based on manipulation of dc motor current, and usually utilizes a current amplifier in the motor's drive unit. The problem with this type of system is the need to have an accurate estimate of the moment of inertia at each joint of the robot arm in order to obtain the desired trajectory. If the actual value of the inertia is smaller than expected, then the torque applied is larger than required. This torque is translated to higher acceleration and consequently higher velocity. This can have disastrous consequences; for example, a part can be struck and broken since the velocity is not zero as desired at the target position. On the other hand, if the inertia is larger than expected there is a loss of time, since the arm decelerates a long distance before the target point and "creeps" toward it very slowly.

An important advantage of the torque-control approach is that we can maintain a desired torque or force. This is useful in some robotics

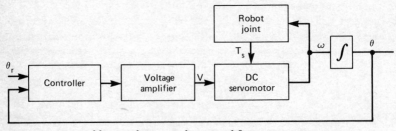

FIG. 3-6   Control loop utilizing a voltage amplifier.

applications, such as screwing or assembly of mating parts. Another advantage is that when the robot arm encounters resistance (e.g., the gripper touches a rigid obstacle) it maintains a constant torque and does not try to draw additional power from the electrical source.

The alternative approach is to control the velocity of the robot arm by manipulation of the dc motor voltage, utilizing a voltage amplifier in the motor's drive unit. A similar approach is also usually used in hydraulically driven robots. The main advantage of this approach is that variations in the moment of inertia affect only the time constant of the response but do not result in any disastrous consequences and do not affect the time required to reach the target position. The arm always approaches the target smoothly and at very low speed. The problem with this approach is that the torque is not controlled, and the motor will draw from the voltage amplifier whatever current is required to overcome the disturbance torque. This can lead to burning of the amplifier's fuse when the robot arm encounters a rigid obstacle. Another disadvantage is that this system is not suitable for certain assembly tasks, such as press fitting and screwing, which require a constant torque or force.

The selected control approach should be dependent on the application and the environment in which the robot arm operates. When the arm is free to move along some coordinate (e.g., spray-painting robots), the specification of velocity is appropriate. When the robot's end effector might be in contact with another object in such a way as to prevent motion along a coordinate, then the specification of torque is appropriate. Note that either velocity or torque may be specified, but not both.

The computer output in velocity-controlled robots can be transmitted either as a sequence of reference pulses or as a binary word in a sampled-data system. With the first technique, the computer produces a sequence of reference pulses for each axis of motion, each pulse generating a motion of 1 BRU of axis travel. (See Chap. 1 for definition of BRU.) The number of pulses represents position, and the pulse frequency is proportional to the axis velocity. These pulses can actuate a stepping motor in an open-loop system or be fed as a reference to a closed-loop robot system. The reference pulse technique can be used only when the velocity control approach is applied to the robot arm.

Most modern CP robots use sampled-data control systems. The sampled-data technique can be used with both the velocity-control and the torque-control approach. With the sampled-data technique, the control loop is closed either through the robot computer (see Fig. 2-8) or in microprocessors, as shown in Fig. 3-7. The hierarchical structure shown in Fig. 3-7 is typical of many modern robot systems. At the top of the system hierarchy is the robot supervisory computer and at the

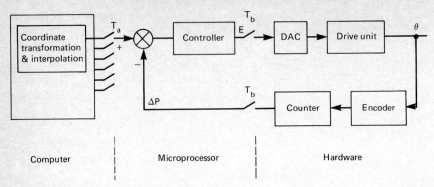

FIG. 3-7   Sampled-data robot system.

lower level are microprocessors—one for each axis of motion. The supervisory computer performs the following functions:

1. Trajectory planning and interpolation in world coordinates.
2. Coordinate system transformation from world to joint coordinates and sending incremental position commands to the microprocessor of each joint every T_a milliseconds (e.g., $T_a = 30$ ms).
3. Receiving a signal from each microprocessor that the corresponding axis has completed its motion and the next end-point coordinate transformation should be performed.
4. In robot systems which include programming with a high-level language, the supervisory computer also contains the language compiler, and the task programming is executed with the aid of this computer.

At the lower level in the system hierarchy is the axial microprocessor. With this structure, the controller block in Figs. 3-5 and 3-6 is included in the microprocessor which controls the corresponding axis. It computes the error signal and sends it through the digital-to-analog converter (DAC) to the drive unit.

In many robot systems, there are two servo loops for each joint control. The inner loop is of an analog type and is closed in the drive unit. It consists of the power amplifier, i.e., the voltage or current amplifier; the joint drive (e.g., a dc servomotor), and a dc tachometer as a velocity feedback device. Further details of this loop are given in Secs. 3.5 and 3.6.

The outer loop is of a sampled-data type. Its typical feedback device is an incremental encoder interfaced with the microprocessor through a counter which is incremented by the pulses received from the encoder. The microprocessor samples the contents of the counter at fixed time intervals T_b. The number transferred from the counter to the computer, ΔP, is equal to the incremental displacement in BRUs. The control program compares the reference from the supervisory computer with the contents of the counter to determine the position error. When the velocity control approach is applied, this error signal is fed every T_b milliseconds (e.g., $T_b = 1$ ms) to a DAC, which, in turn, supplies a voltage proportional to the required axis velocity. When the torque control approach is applied, the error signal is converted in the microprocessor to a corresponding torque signal, and subsequently the latter is used to drive the motor.

The main functions of the axial microprocessor include:

1. Every T_a milliseconds receive a position and/or velocity command from the supervisory computer.
2. Every T_b milliseconds read the counter which stores the position value from the joint encoder.
3. Check if the actual position has reached the required end-point position. If reached, send an appropriate signal to the supervisory computer.
4. Based upon the instantaneous desired and actual positions, calculate the axial position error and use it in a specified control algorithm which depends on the control approach.
5. Send the algorithm result to the DAC.

The DAC converts the binary-word input to a proportional voltage which is used to drive the axis of motion through the drive unit.

The present chapter is concerned with the lower level in the system hierarchy, namely in the axial control loops. The following block diagrams do not show explicitly the encoder, counter, and DAC, but they are always contained in the loops. The analysis and the block diagrams treat the axial loops as a continuous time system, and Laplace transform technique is used to simplify the analysis. This neglects the effect of sampling in the control loop. This is allowed since the sampling period is much smaller than the dominant time constant of the control loop.

3.4 CONTROL LOOPS USING CURRENT AMPLIFIER

One approach to the control of robot joint motions is to apply an appropriate torque to overcome gravity, friction, and dynamic torques due to the moment of inertia J. Several such control loops are discussed in

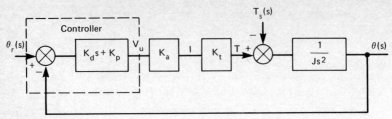

FIG. 3-8 Block diagram of a control loop utilizing a current amplifier.

the literature (Dobrotin & Lewis, 1977; Gill, Paul, & Scheinman, 1974; Lee, 1985; Lee, Mudge, & Turney, 1982; Luh, Walker, & Paul, 1980; Markiewicz, 1973; Paul, 1972; Takase, 1979). A basic control loop with a proportional-derivative (PD) controller and a loop with a compensation for the moment of inertia and gravity torque are discussed in this chapter. Since the motor's armature current is proportional to the loading torque, a current amplifier should be used in torque-control loops. A current amplifier is a device which supplies a current proportional to its input voltage and has a high output resistance.

Basic Loop. A block diagram of the basic control loop using a current amplifier is shown in Fig. 3-5, and its Laplace representation in Fig. 3-8. The model of the motor and the load is based on Eqs. (3-1) and (3-5), which are repeated here using Laplace notation.

$$T = K_t I \tag{3-13}$$

$$\omega(s) = \frac{T - T_s(s)}{sJ} \tag{3-14}$$

Since the position is obtained by integration of the speed

$$\theta(s) = \frac{\omega(s)}{s} \tag{3-15}$$

the amplifier produces current proportional to its input voltage V_u

$$I = K_a V_u \tag{3-16}$$

The typical controller is of a PD type given by

$$V_u(s) = (K_d s + K_p)[\theta_r(s) - \theta(s)] \tag{3-17}$$

where the proportional gain K_p causes a finite steady-state error (for a step input) and the derivative gain K_d must be added for stability considerations. The closed-loop equation is obtained by combining Eqs. (3-13) through (3-17)

$$\theta(s) = \frac{(K_1 s + K_2)\theta_r(s) - T_s(s)}{Js^2 + K_1 s + K_2} \qquad (3\text{-}18)$$

where $K_1 = K_a K_t K_d$ and $K_2 = K_a K_t K_p$.

Note that θ_r and T_s are the Laplace transforms of the required position and static torque, respectively. From Eq. (3-18) it is seen that the closed loop behaves as a second-order system with the characteristic equation

$$s^2 + 2\zeta\omega_n s + \omega_n^2 = 0 \qquad (3\text{-}19)$$

where the damping factor is

$$\zeta = \frac{K_1}{2\sqrt{JK_2}} \qquad (3\text{-}20)$$

and the natural frequency is

$$\omega_n = \sqrt{\frac{K_2}{J}} \qquad (3\text{-}21)$$

The response of the basic loop to a desired position step θ_d and a zero loading torque is calculated by substituting $\theta_r(s) = \theta_d/s$ and $T_s(s) = 0$ in Eq. (3-18), which yields

1. For $\zeta < 1$

$$\theta(t) = [1 - \frac{e^{-\zeta\omega_n t}}{\sqrt{1 - \zeta^2}} \sin{(\sqrt{1 - \zeta^2}\,\omega_n t + \psi)}]\theta_d \qquad (3\text{-}22)$$

where

$$\psi = \tan^{-1} - \frac{\sqrt{1 - \zeta^2}}{\zeta} \qquad (3\text{-}23)$$

2. For $\zeta = 1$

$$\theta(t) = [1 - (1 - \omega_n t)e^{-\omega_n t}]\theta_d \qquad (3\text{-}24)$$

3. For $\zeta > 1$

$$\theta(t) = [1 - \frac{1}{\alpha_1 - \alpha_2}(\alpha_1 e^{-\alpha_1 t} - \alpha_2 e^{-\alpha_2 t})]\theta_d \qquad (3\text{-}25)$$

where

$$\alpha_{1,2} = (\zeta \pm \sqrt{\zeta^2 - 1})\omega_n \qquad (3\text{-}26)$$

Notice that Eqs. (3-22) through (3-25) do not express a typical response of a second-order system. The difference is caused by the term K_1s in the numerator of Eq. (3-18), which does not appear in typical second-order systems.

One drawback of the basic loop is that its damping factor [see Eq. (3-20)] depends on the axial moment of inertia J, which in turn varies with the arm position and the robot payload. For a well-designed manipulator a no-load to full-load variation of inertia of 10:1 can be expected (Paul, 1981). A recommendation given in Gill, Paul, and Scheinman (1974) and Paul (1972, 1981) is to adjust the loop to its critically damped value, namely $\zeta = 1$. This recommendation, however, cannot be satisfied, since the damping factor of each individual loop depends on the manipulator load and the position of the other joints.

EXAMPLE 3-2

A control loop was adjusted to operate with the critically damped value for an average value of $J = J_c$. What is the damping factor range if the joint inertia changes from three times less than the average through three times more than the average?

Solution.   From Eq. (3-20):

$$\frac{\zeta_c}{\zeta} = \sqrt{\frac{J_c}{J}}$$

Since $\zeta_c = 1$, the damping factor for the smallest inertia is 1.73, and for the largest inertia is 0.58.

The responses to a step position input with the damping factors obtained in Example 3-2 are demonstrated in Fig. 3-9. The natural frequency for the average value of J is denoted by ω_{n0}, and consequently the natural frequency for the minimum J is $1.73\omega_{n0}$, and for the maximum J is $0.58\omega_{n0}$. Therefore, a smaller inertia causes a faster response together with a larger damping factor, and thus a decrease in the maximum overshoot. Notice, however, that with this loop an overshoot always exists [caused by the term K_1s in the numerator of Eq. (3-18)]. For example, for $\zeta = 1$ the overshoot is 13.5 percent, and for $\zeta = 1.73$ it is 6 percent. The maximum allowable damping factor depends on the maximum allowable current and acceleration of the axial motor.

The objective of the basic torque loop is to deliver the torque required to drive the corresponding joint. Substituting Eq. (3-18) into (3-17) and combining with Eqs. (3-13) and (3-16) yields the Laplace transform of the torque produced by the motor

$$T(s) = \frac{(K_1s + K_2)[Js^2\theta_r(s) + T_s(s)]}{Js^2 + K_1s + K_2} \qquad (3\text{-}27)$$

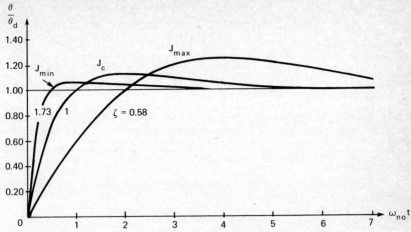

FIG. 3-9   Position response to a step input in a basic torque-control loop.

The term T_s in Eq. (3-27) contains two types of torques [see Eq. (3-5)]. One is caused by the coupling inertia and centripetal torques and therefore disappears when the arm is stationary, and the other is a static torque which is mainly caused by gravity (T_g). Therefore, at the steady state $T_s = T_g/s$ and by using the final-value theorem [given in Eq. (3-53)], Eq. (3-27) yields $T = T_g$, which means that when the motor is stationary it produces the exact torque required to overcome the gravity forces. However, the required dynamic torque $Js^2\theta_r$ is not achieved, since the left-hand part of the numerator is not equal to the denominator in Eq. (3-27).

One disadvantage of this control loop is the existence of a position error at the steady state. This error is calculated from Eq. (3-18) for $s = 0$ and $T_s = T_g$

$$E = \theta_r - \theta = \frac{T_g}{K_2} \tag{3-28}$$

The control program must contain an algorithm to compensate for this error.

Torque Loop with Compensations.   In order to improve the performance of the control loop discussed above, the controller algorithm must compensate for the joint inertia, the gravity, and the required dynamic torque. Consequently, it is modified as follows (see Fig. 3-10):

1. An estimation of the moment of inertia $\hat{J}$ is inserted as a programmed gain. The gain $\hat{J}$ can be introduced either as in Fig. 3-10 (Paul, 1972)

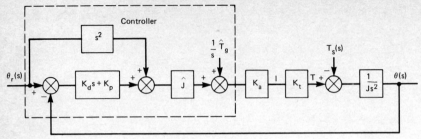

FIG. 3-10   Compensated torque-control loop.

or in the acceleration feedforward block alone, i.e., $\hat{J}s^2$ (Markiewicz, 1973; Paul, 1981).

2. An estimation of the static torque due to gravity T_g is programmed in order to reduce (or eliminate) the steady-state error in Eq. (3-28).

3. An acceleration feedforward term is added in order to improve the accuracy in obtaining the required dynamic torque.

The equations of the compensated loop are as follows:

$$T = \left\{ [(\theta_r - \theta)(K_d s + K_p) + s^2\theta_r]\hat{J} + \frac{\hat{T}_g}{s} \right\} K_a K_t \qquad (3\text{-}29)$$

and

$$\theta = \frac{T - T_s}{Js^2} \qquad (3\text{-}30)$$

Combining Eqs. (3-29) and (3-30) yields the closed-loop equation

$$\theta = \frac{\hat{J}(K_a K_t s^2 + K_1 s + K_2)\theta_r - T_s + K_a K_t \hat{T}_g/s}{Js^2 + \hat{J}K_1 s + \hat{J}K_2} \qquad (3\text{-}31)$$

where $K_1 = K_a K_t K_d$ and $K_2 = K_a K_t K_p$.

Equation (3-31) represents a second-order system with a damping factor of

$$\zeta = \frac{K_1}{2} \sqrt{\frac{\hat{J}}{K_2 J}} \qquad (3\text{-}32)$$

and a natural frequency of

$$\omega_n = \sqrt{\frac{\hat{J}K_2}{J}} \qquad (3\text{-}33)$$

In this system the amplifier gain is adjusted such that

$$K_a K_t = 1 \tag{3-34}$$

Consequently, Eq. (3-31) becomes

$$\theta = \frac{(s^2 + K_1 s + K_2)\theta_r - (T_s - \hat{T}_g/s)/\hat{J}}{(J/\hat{J})s^2 + K_1 s + K_2} \tag{3-35}$$

and the steady-state position error for a step input is

$$E = \theta_r - \theta = \frac{T_s - \hat{T}_g/s}{\hat{J}K_2} \tag{3-36}$$

where T_s is the Laplace transform of a torque which at the steady state is a constant T_g caused by the gravity force, namely

$$T_s = \frac{T_g}{s} \tag{3-37}$$

If the estimated torque $\hat{T}_g$ is equal to the actual torque T_g, the steady-state position error is zero.

Similarly, if the estimated inertia $\hat{J}$ is equal to the actual inertia J, Eq. (3-35) yields the ideal situation $\theta = \theta_r$ and consequently, from Eq. (3-30)

$$T = J s^2 \theta_r + \frac{T_g}{s} \tag{3-38}$$

so that the motor always produces the required dynamic and static torques.

The obvious problem with this type of system is the need to have an accurate estimate of the changing gravity torque and moment of inertia J in order to obtain the desired position and dynamic response. In order to find the response to a step input let us assume that $T_s = \hat{T}_g/s$. Substituting $\theta_r = \theta_d/s$ and $c = \hat{J}/J$ into Eq. (3-35) and obtaining the inverse Laplace transform yields the following results:

1. For $\zeta < 1$

$$\theta(t) = \left[1 - \frac{1 - c}{\sqrt{1 - \zeta^2}} e^{-\zeta \omega_n t} \sin\left(\sqrt{1 - \zeta^2} \omega_n t + \psi \right) \right] \theta_d \tag{3-39}$$

   where ψ is defined in Eq. (3-23)

2. For $\zeta = 1$

$$\theta(t) = [1 - (1 - c)(1 - \omega_n t)e^{-\omega_n t}]\theta_d \tag{3-40}$$

3. For $\zeta > 1$

$$\theta(t) = \left[1 - \frac{1 - c}{\alpha_1 - \alpha_2} (\alpha_1 e^{-\alpha_1 t} - \alpha_2 e^{-\alpha_2 t}) \right] \theta_d \qquad (3\text{-}41)$$

Where α_1 and α_2 are defined in Eq. (3-26). Notice that in all three cases $\theta(0) = c\theta_d$.

Except for the term $(1 - c)$, Eqs. (3-39) through (3-41) are similar to Eqs. (3-22) through (3-25) of the basic loop. If the moment of inertia is well estimated, then $c = 1$ and the ideal response $\theta(t) = \theta_d$ is obtained regardless of ζ. If, however, $\hat{J} \neq J$, catastrophic results might occur. This is demonstrated in Fig. 3-11. Assume that the gain $\hat{J}$ was adjusted so that $\zeta = 0.71$ for $\hat{J} = J_{av}$ and consequently the corresponding response is an ideal step. If the actual J becomes eight times larger, then the damping factor is reduced to $0.71/\sqrt{8} = 0.25$, which results in an overshoot of 46 percent. A similar phenomenon occurs also with the basic loop. However, if the actual J becomes eight times smaller than $\hat{J}$, then $\theta(0) = 8\theta_d$ and the corresponding overshoot is 700 percent! This must be avoided at all costs.

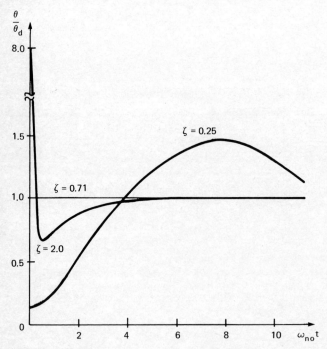

FIG. 3-11 Position response of a misadjusted compensated torque loop.

The compensated loop can operate either with a variable gain where $\hat{J}$ varies during the arm motion or with a fixed-gain $\hat{J}$. If $\hat{J}$ is a fixed gain, the best performance is obtained by adjusting it to $\hat{J} = J_{min}$. This guarantees that $c \leq 1$. When the actual inertia is at its minimum, the maximum damping factor is achieved, as can be seen from Eq. (3-32). The minimum damping factor occurs at J_{max}, so to avoid large overshoots it is desirable to adjust the gains such that $\zeta_{min} > 1$. This min-max adjustment method produces

$$c = \frac{J_{min}}{J} \leq 1$$

$$\zeta_{min} = \frac{K_1}{2} \sqrt{\frac{J_{min}}{K_2 J_{max}}} > 1$$

and

$$\zeta = \zeta_{min} \sqrt{\frac{J_{max}}{J}}$$

Figure 3-12 demonstrates responses obtained by this method when variations of 10:1 are expected in the effective inertia. The minimum

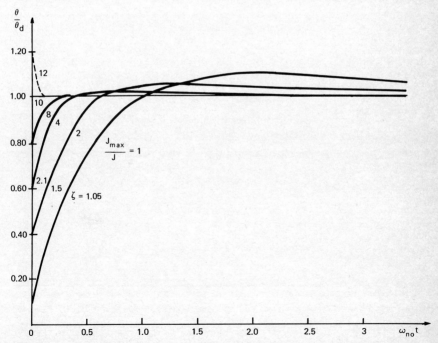

FIG. 3-12   Position response in a compensated loop adjusted according to the min-max method.

damping factor is adjusted to $\zeta = 1.05$, and results in a maximum overshoot of 11 percent. For a smaller inertia the overshoot is smaller as well. For $J = J_{min} = 0.1J_{max}$, the response is an ideal step. However, if for any reason $J < J_{min}$ an overshoot occurs at $t = 0$, as is shown by the dashed line in Fig. 3-12.

The compensated loop might provide a satisfactory solution for variable-gain loops, in which the value of $\hat{J}$ is continuously adjusted by the robot computer. In practice, however, commercial robots operate with fixed-gain loops, and in these cases the compensated loop has the following drawbacks:

1. There always exists an overshoot to a step response. This situation can be remedied if a tachometer feedback is added to the control loop. In this case, however, the loop is no longer a torque-control loop.

2. The double derivative (s^2) does not function properly when step or ramp inputs are provided. The above analysis assumed a linear model, and consequently the response of a derivative to a step or a double derivative to a ramp input is an infinite impulse. But the allowable current to the motor is limited. This means that during the initial starting period the response of the motor is

$$\theta = \frac{K_t I_m t^2}{2J} \tag{3-42}$$

regardless of the value of $\hat{J}$ or θ_d. This response continues until the current is reduced below its maximum value (I_m) by the feedback.

3. Errors due to approximations in modeling (e.g., $T = \hat{J}\ddot{\theta} + T_s$) and system nonlinearities prevent the ideal response for $\hat{J} \approx J$ and the prediction of the exact response in other cases.

4. Gravity torques must be computed in real time in order to be compensated. This requires a large program and a lengthy computing time (Markiewicz, 1973).

3.5 CONTROL LOOP USING VOLTAGE AMPLIFIER

An alternative approach is to control the speed of the robot joint by manipulation of the motor voltage utilizing a voltage amplifier, as shown in Fig. 3-6. A voltage amplifier provides an output voltage proportional to its input voltage and is capable of supplying the current required by the motor.

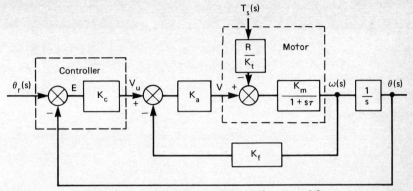

FIG. 3-13   Block diagram of control loop utilizing voltage amplifier.

A block diagram of a basic control loop is shown in Fig. 3-13. The output of the loop is defined as either the speed or the position of the robot joint. The torque T_s is mainly due to coupling inertia and gravity acts as a disturbance on the motor.

The control loop in Fig. 3-13 includes an inner loop consisting of the voltage amplifier with a gain K_a, the dc motor, and a tachometer as a velocity feedback device with a gain K_f. This loop is frequently sold as a package and therefore is denoted as the drive unit in Fig. 3-7. The transfer function of the drive unit is derived as follows.

The input voltage to the motor is

$$V(s) = K_a[V_u(s) - K_f\omega(s)] \qquad (3\text{-}43)$$

Combining the motor's speed equation, Eq. (3-9), with Eq. (3-43) yields

$$\omega(s) = \frac{\alpha K_a K_m V_u(s) - (RK_m/K_t)\alpha T_s(s)}{1 + s\alpha\tau} \qquad (3\text{-}44)$$

where we have defined an attenuation factor

$$\alpha = \frac{1}{1 + K_a K_f K_m} \qquad (3\text{-}45)$$

Comparison of Eq. (3-9) with (3-44) shows that the forms of the system equations are the same, and that the effect of the tachometer feedback is to reduce the time constant (since $\alpha < 1$, then $\alpha\tau < \tau$), to reduce the effect of the load torque, to reduce any nonlinearities of the voltage amplifier, and to facilitate the adjustment of the overall gain by adjusting the gain K_f (Koren, 1983).

Comparing the loop structure of Fig. 3-8 and Fig. 3-13 shows that the derivative controller is no longer necessary and a proportional controller, with a gain K_c, is sufficient. The closed-loop equation is obtained by combining Eq. (3-44) with (3-15) and the controller equations

$$E(s) = \theta_r(s) - \theta(s) \tag{3-46}$$

and

$$V_u(s) = K_c E(s) \tag{3-47}$$

which yields

$$\theta(s) = \frac{K\theta_r(s) - K_q T_s(s)}{\tau' s^2 + s + K} \tag{3-48}$$

where K is the open-loop gain defined by

$$K = \alpha K_a K_m K_c \tag{3-49}$$

K_q is a gain defined by

$$K_q = \frac{\alpha R K_m}{K_t} $$

and

$$\tau' = \alpha\tau = \frac{\alpha R J}{K_t K_v} \tag{3-50}$$

The characteristic equation of the closed loop is given in Eq. (3-19) where the damping factor is

$$\zeta = \frac{1}{2\sqrt{K\tau'}} \tag{3-51}$$

and the natural frequency is

$$\omega_n = \sqrt{\frac{K}{\tau'}} \tag{3-52}$$

The actual position response to a position step input in a critically damped system ($\zeta = 1$) is shown in Fig. 3-14 (for $T_s = 0$). Note that in this case the overshoot is zero, compared with 11 percent in Fig. 3-12. The problem is, however, that since τ' is proportional to the inertia J, the dependence on J in Eqs. (3-51) and (3-52) is similar to that in Eqs. (3-20) and (3-21) in the loop utilizing a current amplifier. As a consequence, the present loop also has the unfavorable

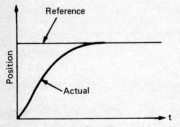

FIG. 3-14 Required and actual position response in a speed control loop.

situation of a damping factor which depends on a changing moment of inertia. Similarly, this loop has not remedied the other problem of the basic loop utilizing a current amplifier, namely, the existence of a torque-dependent position error at the steady state [see Eq. (3-28)].

To conclude, the characteristic equation of the two systems representing control utilizing a current amplifier and control with the voltage amplifier is similar. When the power amplifier is saturated, the resulting nonlinear behavior of the two systems is different. The analysis of a nonlinear behavior, however, is beyond the scope of this text.

3.6 ELIMINATION OF STATIONARY POSITION ERRORS

The steady-state position error of the control loop shown in Fig. 3-13 is derived by substituting Eqs. (3-37) and (3-48) into (3-46) and using the final-value theorem

$$\lim_{s \to 0} sF(s) = \lim_{t \to \infty} f(t) \tag{3-53}$$

which yields

$$E = \frac{K_q T_g}{K} \tag{3-54}$$

This is a position error due to gravity forces that exist at the end point. The explanation of this error can be found by substituting the values K_q and K from Eq. (3-49) into (3-54) which yields

$$EK_a K_c = \frac{R}{K_t} T_g \tag{3-55}$$

Equation (3-55) means that when the joint is stationary the voltage amplifier supplies a voltage $V = EK_a K_c$ to counteract the effect of the gravity torque T_g. To generate this voltage, a position error E must exist and consequently the joint does not reach the required end-point position. In the compensated current-amplifier loop this situation was remedied by programming an estimated gravity torque to counteract the real one. Obviously, the same approach can be also applied here. However, since the real gravity torque depends upon the angle values of the various joints, it is difficult to have an accurate estimate of the torque values for every position of the manipulator, and therefore this method has only low practical usefulness.

An alternative approach to eliminate the stationary (i.e., the steady state) position error is to add an integral or a proportional-integral (PI) controller into the internal loop of Fig. 3-13. The proposed internal

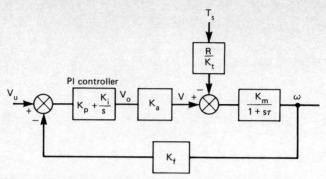

FIG. 3-15   A velocity loop containing an integral controller.

loop with a PI controller is shown in Fig. 3-15. The input to an integrator at steady state must be zero, and therefore with this loop $\omega = V_u = 0$. Since V_u is zero, the steady-state position error E is zero as well, and the joint reaches the desired end position. The output of the PI controller V_o generates the voltage V required to overcome the effect of gravity at steady state.

The proposed control loop requires a careful design since the characteristic equation is of the third order rather than second order as in the previous loop [see Eq. (3-48)], and inappropriate selection of the loop gains will cause an unstable system. An improved stability is obtained by using the PI controller in the internal loop rather than an integral alone. The PI controller guarantees zero position error when the joint is in no motion, together with an unoscillatory response during the motion itself.

3.7 CONTROL LOOPS OF CNC SYSTEMS

The control loops which are utilized in CNC systems of machine tools are used in robotics as well (Koren, 1983). The structure of a typical loop is shown in Fig. 3-16 and is actually similar to the one presented in Sec. 3.5. The difference is in the type of input. The intput to the CNC loop is a velocity command rather than a position command, and consequently the corresponding shaft velocity is declared as the loop output. In order to maintain a zero velocity error at the steady state for a step input, the controller must contain an integration action ($1/s$ in Laplace notation). As a result, the structure of the control loops in Figs. 3-13 and 3-16 is similar and consequently their mathematical model is identical [i.e., Eqs. (3-46) through (3-52) are valid in this case as well].

In addition to the velocity reference ω_r, a position reference θ_r is also

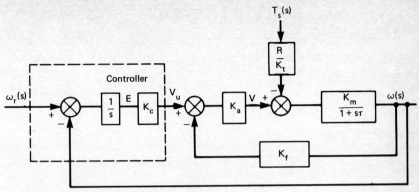

FIG. 3-16   Control loop of a CNC system.

sent by the interpolator of the CNC supervisory computer. Every sampling period this reference is compared with the actual position presented by the pulses from the encoder. When both the reference and actual positions are equal, the joint is at its desired position and the velocity reference must be stopped.

The typical velocity reference in robotic systems is a step input, namely, a constant joint velocity is desired. The position error E_s for a constant velocity ω_r and a constant torque T_s are derived by substituting Eq. (3-48) into (3-46), using the relationship $\omega_r(s) = s\theta_r(s)$ and the final-value theorem

$$E_s = \frac{\omega_r}{K} + \frac{K_q T_s}{K} \tag{3-56}$$

Figure 3-17 shows the typical inputs to the control loop: A velocity step ω_r which is used as a reference to the control loop, and a position increment θ_r, the distance to the next target point (can be either the end point or a point along the trajectory as marked on Fig. 2-6). The velocity step can be converted in the loop controller (see Fig. 3-7) to its counterpart position ramp (shown in the dashed line in Fig. 3-17), which is used in the position calculations.

The velocity reference is stopped at time t_1. The time t_1 is determined by comparing the level of the dashed ramp in Fig. 3-17 to θ_r. At t_1 the difference between the target position and the actual position is E_s [given in Eq. (3-56)], but the control loop corrects this error during the period t_1 through t_2. As a result an automatic deceleration is achieved until the axis completely stops at time t_2.

With this system, although the dependence of the damping factor on J exists, it does not have any critical effect as in the previous systems. The reason is that here the damping factor affects mostly the velocity

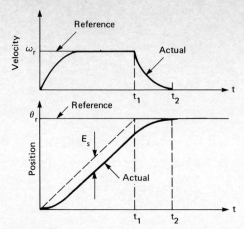

FIG. 3-17   Velocity and position response in a CNC-type control loop.

rather than position response. With the present system a smaller damping factor, for example, increases the level of overshoot in the velocity, which has only a minor effect on the actual position responses in Fig. 3-17. By contrast, a smaller damping factor in the previous systems causes an overshoot above the target position, which might have a disastrous consequence when, for example, the robot has to put an object on a hard surface.

The gain-tuning problem is further explained in what follows. A satisfactory performance is obtained with a CNC-type robot system when the loop gain K is adjusted according to

$$K = \frac{1}{2\tau'} \tag{3-57}$$

where $\tau' = \alpha\tau$; the constant α is given in Eq. (3-45), and τ is the mechanical time constant, given in Eq. (3-8) and is repeated here

$$\tau = \frac{JR}{K_t K_v} \tag{3-58}$$

The moment of inertia J is given by

$$J = J_r + K_g^2 J_l$$

where J_r is the moment of inertia of the motor rotor, J_l is the moment of inertia of the load, and K_g is the gear ratio between the load and the motor shaft. If J_l is time-varying, then τ becomes time-varying as well and the condition given by Eq. (3-57) cannot be satisfied unless K is a variable gain. The damping factor ζ of the loop is given by

$$\zeta = \frac{1}{2\sqrt{K\tau'}} \qquad (3\text{-}59)$$

In order to avoid large overshoots, the condition $\zeta > 0.7$ should be maintained. Therefore, the gain K is tuned for the maximum τ which depends on the maximum expected moment of inertia. However, this causes very large damping factors for small moments of inertia, which results in a very sluggish transient behavior that decreases the overall operating speed.

Finally, in order to eliminate end-point position errors due to gravity, a PI controller must be added into the internal loop of Fig. 3-16. It is worthwhile to note that the PI controller can be eliminated in axes which are not affected by gravity, such as the robot base axis. The combination of a velocity step (or a position ramp) input to the control loop of Fig. 3-16 together with the PI action in the internal loop provides satisfactory results at the target positions.

3.8 CONCLUSIONS AND ASSESSMENTS

The maximum traveling velocity of the manipulator, the amount of overshoot, and settling time at the target point are the dominant dynamic parameters in robotics. The traveling velocity, permittable acceleration, and settling time provide the overall speed of operation, and the amount of overshoot can change the shape of the generated path or cause disastrous collisions in assembly when the tool collides with an obstruction.

In robots which contain a rotary base, a good dynamic performance is usually difficult to achieve. The effective moment of inertia at the base depends not only on the weight of the object being carried but also upon the instantaneous velocity and position of the end effector and the other joints during the motion. As a result, the moment of inertia reflected at the base is in general time-varying. A similar problem exists with other rotary joints.

As a result, the open-loop gains of rotary axes cannot be adjusted to obtain optimal dynamic performance. Each loop gain in the controller must be tuned for a certain inertia load in order to avoid overshoots over the target point. This tuning degrades the performance when other moment of inertia loads are present.

The problem of the gain dependence on J motivated research in adaptive control systems for robotics. These systems are based upon real-time estimation of *J*, and a subsequent adjustment of the loop gain to meet the desired performance. The computing processes of estimation and gain adjustment must be performed at the beginning of each motion between end points. The main problem with adaptive control systems for robots is that the typical time duration of motions is too short for performing the two processes.

Another experimental control method is the resolved-acceleration control (Luh, Walker, and Paul, 1980). This method deals directly with the position and orientation of the hand. It differs from others in that accelerations are specified and that all the feedback control is done at the hand level. However, neither this nor the adaptive-control algorithm is sufficiently mature to be applied in commercial robots. The most widely used method today applies a separate control loop for each joint designed with linear-control laws, as was widely discussed in this chapter.

REFERENCES

Cvetkovic, V., and M. Vukobratoric: "Contribution to Controlling Non-Redundant Manipulators," *Proc. 3rd Symp. Theory Pract. Robots Manip.*, Udine, September 1978.

Dobrotin, B., and R. A. Lewis: "A Practical Manipulator System." *Proc. 5th Int. Conf. Artif. Intell.*, MIT, Cambridge, Mass., August 1977, pp. 723–732.

Gill A., R. Paul, and V. Scheinman: "Computer Manipulator Control, Visual Feedback and Related Problems," *Proc. 1st Symp. Theory Prac. Robots Manip.*, Udine, September 1974, pp. 31–50.

Koren, Y.: *Computer Control of Manufacturing Systems*, McGraw-Hill Book Co., New York, 1983.

Koren, Y., and G. A. Ulsoy: "Control of DC Servomotor Driven Robots," *Proc. Robot VI Conf.*, Detroit, Mich., March 1982, pp. 590–602.

Lee, C. S. G.: *Robotics Theory and Practice.* Addison-Wesley, Reading, Mass., to be published.

Lee, C. S. G., T. N. Mudge, and J. L. Turney: "Hierarchical Control Structure Using Special Purpose Processors for the Control of Robot Arms," *Proc. IEEE Conf. Pat. Recog. Image Process.*, Las Vegas, Nev., June 1982.

Luh J. Y., M. W. Walker, and R. P. Paul: "Resolved-Acceleration Control of Mechanical Manipulators," *IEEE Trans. Autom. Control*, vol. 25, no. 3, 1980, pp. 468–474.

Markiewicz, B. R.: "Analysis of Computed Torque Drive Method and Comparison with Conventional Position Servo for a Computer-Controlled Manipulator," *NASA Tech. Memo 33-601.* J.P.L., March 1973.

Paul, R. P.: *Robot Manipulators: Mathematics, Programming, and Control*, MIT Press, Cambridge, Mass., 1981.

Paul, R. P.: *Modelling, Trajectory Calculation and Serving of a Computer Controlled Arm*, Ph.D. Thesis, Stanford University, November 1972. (Also Stanf. Artif. Intell. Memo 147, March 1973).

Paul, R. P.: "Manipulator Cartesian–Path Control." *IEEE Trans. Syst., Man. Cybern.*, vol. SMC-7, no. 11, November 1979, pp. 702–711.

Saridis, G. N., and C. S. G. Lee: "Heuristic Control in Trainable Manipulators," *Proc. Joint Autom. Control Conf.*, San Francisco, Cal., 1976, pp. 712–716.

Saridis, G. N., and C. S. G. Lee: "An Approximation Theory of Optimal Control for Trainable Manipulators," *IEEE Trans. Syst., Man., Cybern.*, vol. SMC-9, no. 3, March 1979, pp. 152–159.

Scheinman, V. D.: "Design of a Computer Manipulator," Stanf. Artif. Intell. Memo 92, June 1969.

Takase, K.: "Skill of Intelligent Robot," *Proc. 6th Int. Conf. Artif. Intell.*, Tokyo, August 1979, pp. 1095–1100.

Whitney, D. E.: "Force Feedback Control of Manipulators' Fine Motions," *ASME Trans. J. Dyn. Syst. Meas. Control*, June 1977, pp. 91–97.

Yuan, J. S. C.: "Dynamic Decoupling of a Remote Manipulator System" *IEEE Trans. Autom. Control.*, vol. AC-23, no. 4, August 1978, pp. 713–717.

CHAPTER 4

Kinematic Analysis and Coordinate Transformation†

In robotics, the solution of the direct kinematics problem involves the determination of the end effectors' position and orientation and their rate of change, as a function of the given positions and speeds of the axes of motion, as shown in Fig. 4-1. The position of the end effector, or robot tool, is defined at the TCP, which is, for example, the edge of a welding gun or the center of a gripped object (see Sec. 1.5).

A demonstrative example of direct kinematic analysis using straight-forward geometric solution is discussed in Sec. 4.2. This method is suitable for simple kinematic structures but becomes difficult to apply when dealing with multiple-degree-of-freedom mechanisms.

In Sec. 4.3 through 4.5 a method based on transformation matrices will be presented. Compared with the previous method, this method is suitable for analyzing multiple-degree-of-freedom mechanisms and is also ultilized for some of the inverse kinematics techniques and manipulator dynamic analysis.

Another general method, using rotation matrices, is presented in Sec. 4.6. Applying this method to solve the inverse kinematics problem is preferable in several cases, as will be shown in Chap. 5.

4.1 DIRECT KINEMATICS PROBLEM IN ROBOTICS

The kinematic analysis of a mechanical system means the determination of the position, velocity, and acceleration of the various mechanical ele-

†This chapter was contributed by G. Amitai and M. Shoham.

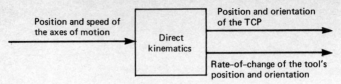

FIG. 4-1   The direct kinematics transformation.

ments forming the mechanism under consideration. The combination of position and velocity of an element at a certain time is referred to henceforth as the *state* of this element. The effect of the associated forces and torques, which take into account the mass and inertia of the mechanical elements, is considered in dynamic rather than kinematic analysis, and therefore is not discussed in this chapter. The calculation of the kinematic state of mechanical elements (i.e., their position, orientation, and rate of change) on the basis of given axial motions is referred to as the *direct kinematics problem.* The *inverse kinematics problem* refers to the calculation of required axis motions to produce desired arm motions.

In the case of manipulators, direct kinematic analysis determines the state of the end effector as a function of the known states of the various joints. In addition to the state of the end effector, direct kinematic analysis also includes the determination of the links' position and orientation and their time derivatives. Such information is essential for any subsequent dynamic analysis or when the position and orientation of a sensor mounted on a link are required for the data processing.

In this chapter several analytical techniques for the solution of the direct kinematics problem will be presented. These techniques can be ultilized not only for the analysis of motion of the manipulator but also in other motion-related robotic problems, such as transformation of coordinates. For example, the coordinate system attached to a sensor can be transformed to the world coordinate system (WCS) assigned to the robot's stationary base. Another example is that the tool coordinate system (TCS), which is attached to the end effector, can be transformed to any other coordinate system. The transformed information can be a desired velocity vector defined in the TCS or a position vector in the sensor coordinate system (SCS) (see Fig. 5-1).

The direct kinematic analysis provides the basis for solving the inverse kinematics problem. The solution of the inverse kinematics problem becomes the basis of the interpolator algorithm which guides the end effector along the desired trajectory in space. The interpolator supplies reference signals to the control loops which, in turn, cause the end effector to follow the desired trajectory.

The direct kinematics and the inverse kinematics solutions are implemented in path-planning algorithms of CP robots. Basically, direct kinematics is used to transform trajectory points given in joint coordinate system (JCS) or TCS to WCS, and inverse kinematics is used to transform the given trajectory points into axial motions.

This chapter will be concerned with simple joints (where a one-degree-of-freedom linkage forms a joint) but not with complex joints.

4.2 GEOMETRY-BASED DIRECT KINEMATIC ANALYSIS

A two-degree-of-freedom articulated planar mechanism is shown in Fig. 4-2a. Link 2 rotates around joint 2 at an angle θ_2 measured with respect to a reference coordinate system (X_1, Y_1) which is attached to link 1. Likewise, link 3 rotates around joint 3 at an angle θ_3 with respect to link 2. The TCP is defined at the tip of link 3.

The direct kinematics problem of defining the position of the TCP (X_t, Y_t) as a function of the joint angle values can be solved using straightforward geometry

$$X_t = a_2 \cos \theta_2 + a_3 \cos \psi$$
$$Y_t = a_2 \sin \theta_2 + a_3 \sin \psi \tag{4-1}$$

where $\psi = \theta_2 + \theta_3$. The time derivative of Eq. (4-1) gives the velocity of the TCP in the reference coordinate system (X_1, Y_1):

$$\dot{X}_t = - [a_2 \sin \theta_2 + a_3 \sin \psi]\dot{\theta}_2 - [a_3 \sin \psi]\dot{\theta}_3$$
$$\dot{Y}_t = [a_2 \cos \theta_2 + a_3 \cos \psi]\dot{\theta}_2 + [a_3 \cos \psi]\dot{\theta}_3 \tag{4-2}$$

Equations (4-1) and (4-2) represent the direct kinematics solution for the mechanism shown in Fig. 4-2a.

A similar solution can be found for the three-degree-of-freedom spherical mechanism shown in Fig. 4-2b. In this case the reference coordinate system (X_0, Y_0, Z_0) is attached to the stationary base, which is also referred to as link O. The position of the TCP, expressed in the base coordinate system, is

$$X_t = d_3 \cos \theta_1 \cos \theta_2$$

$$Y_t = d_3 \sin \theta_1 \cos \theta_2 \tag{4-3}$$

$$Z_t = d_3 \sin \theta_2$$

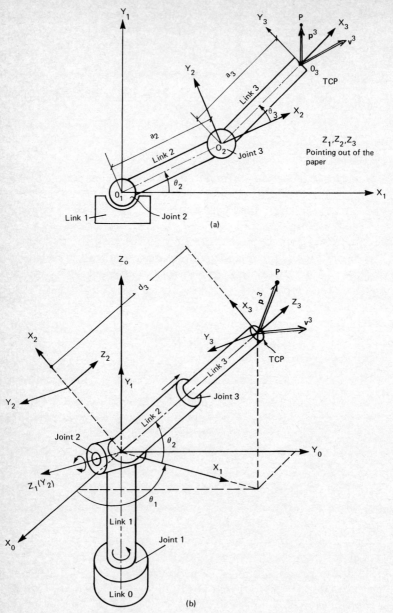

FIG. 4-2 Robot manipulators; the assignment of links' coordinate systems according to the DH method (Sec. 4.5.1).

and the velocity of the TCP is given by the time derivative of Eq. (4-3):

$$\dot{X}_t = (\cos\theta_1 \cos\theta_2)\dot{d}_3 - (d_3 \sin\theta_1 \cos\theta_2)\dot{\theta}_1 - (d_3 \cos\theta_1 \sin\theta_2)\dot{\theta}_2$$

$$\dot{Y}_t = (\sin\theta_1 \cos\theta_2)\dot{d}_3 + (d_3 \cos\theta_1 \cos\theta_2)\dot{\theta}_1 - (d_3 \sin\theta_1 \sin\theta_2)^{\dot{\theta}_2} \quad (4\text{-}4)$$

$$\dot{Z}_t = (\sin\theta_2)\dot{d}_3 + (d_3 \cos\theta_2)\dot{\theta}_2$$

Again, Eqs. (4-3) and (4-4) are the direct kinematics solution for the mechanism in Fig. 4-2b.

Another example of a three-degree-of-freedom articulated open kinematic chain mechanism can be a mechanism constructed by replacing links 2 and 3 in Fig. 4-2b by links 2 and 3 from Fig. 4-2a (the prismatic joint 3 is replaced by a revolute joint). The position of the TCP in this case is given by

$$X_t = a_2 \cos\theta_2 \cos\theta_1 + a_3 \cos\psi \cos\theta_1$$

$$Y_t = a_2 \cos\theta_2 \sin\theta_1 + a_3 \cos\psi \sin\theta_1 \quad (4\text{-}5)$$

$$Z_t = a_2 \sin\theta_2 + a_3 \sin\psi$$

and the velocity of the TCP is given by

$$\dot{X}_t = -(a_2 \cos\theta_2 \sin\theta_1 + a_3 \cos\psi \sin\theta_1)\dot{\theta}_1$$

$$-(a_2 \sin\theta_2 \cos\theta_1 + a_3 \sin\psi \cos\theta_1)\dot{\theta}_2$$

$$-(a_3 \sin\psi \cos\theta_1)\dot{\theta}_3$$

$$\dot{Y}_t = (a_2 \cos\theta_2 \cos\theta_1 + a_3 \cos\psi \cos\theta_1)\dot{\theta}_1 \quad (4\text{-}6)$$

$$-(a_2 \sin\theta_2 \sin\theta_1 + a_3 \sin\psi \sin\theta_1)\dot{\theta}_2$$

$$-(a_3 \sin\psi \sin\theta_1)\dot{\theta}_3$$

$$\dot{Z}_t = (a_2 \cos\theta_2 + a_3 \cos\psi)\dot{\theta}_2 + (a_3 \cos\psi)\dot{\theta}_3$$

where $\psi = \theta_2 + \theta_3$.

Observing the three examples, one can see that there is no systematic approach applied and the mathematical complexity depends on the individual kinematic structure. Moreover, it is expected that the complexity of the solution increases with the number of degrees of freedom in the mechanism. For these cases the methods described in the next sections are preferable, since they are almost independent of the complexity of the mechanism and of the number of degrees of freedom.

4.3 COORDINATE AND VECTOR TRANSFORMATIONS USING MATRICES

A transformation in kinematics is the process of setting up correspondences between the elements of two coordinate systems that are translated and/or rotated relative to each other. The location of a point can be expressed in each of these coordinate systems. When vector transformation is considered, the projections of the vector are transformed. The magnitude of the vector, however, is fixed and does not depend on the coordinate system.

There are several transformation techniques:

1. Complex numbers, a technique which is mainly useful in solving two-dimensional cases, but becomes complicated in three-dimensional cases.

2. Rotation matrices, a technique which is useful for vector transformations (e.g., see Sec. 4.3.1).

3. The quaternion and rotation vectors (see Sec. 4.6).

4. The homogeneous transformation matrices technique which is discussed in the present section.

A useful method based upon the homogeneous transformation matrices was established by Denavit and Hartenberg (1955) for use in kinematic chains. This method is discussed in the next section.

The first step in implementing the transformation matrices technique to kinematic chains is to assign coordinate systems to the moving links of the manipulator. A right-handed coordinate system O_i, denoted henceforth as the O_i frame, is attached to link i with the origin at joint $i + 1$. The joints and links are numbered starting at the stationary base, as demonstrated in Fig. 4-2.

The transformation matrices define the geometric relationship between two consecutive coordinate systems O_i and O_{i-1}, as a function of the joint variable Q_i where $Q_i = \theta_i$ for revolute joints and $Q_i = d_i$ for prismatic joints. Each transformation matrix is composed of an orientation matrix and a translation vector, which are defined below.

4.3.1 The Orientation Matrix and Translation Vector

Figure 4-3 represents a planar case where frame O_i is rotated and translated relative to O_{i-1}. The vector components of $\mathbf{v}^{i-1}$ can be expressed by the vector components of $\mathbf{v}^i$.

$$v_x^{i-1} = v_x^i \cos \theta_i - v_y^i \sin \theta_i \tag{4-7}$$

$$v_y^{i-1} = v_x^i \sin \theta_i + v_y^i \cos \theta_i \tag{4-8}$$

Equations (4-7) and (4-8) can be expressed in a matrix form

$$\begin{bmatrix} v_x^{i-1} \\ v_y^{i-1} \end{bmatrix} = \begin{bmatrix} \cos \theta_i & -\sin \theta_i \\ \sin \theta_i & \cos \theta_i \end{bmatrix} \begin{bmatrix} v_x^i \\ v_y^i \end{bmatrix} \tag{4-9}$$

Alternatively, Eq. (4-9) is written as

$$\mathbf{v}^{i-1} = \mathbf{C}_{i-1}^i \mathbf{v}^i \tag{4-10}$$

where

$$\mathbf{C}_{i-1}^i = \begin{bmatrix} \cos \theta_i & -\sin \theta_i \\ \sin \theta_i & \cos \theta_i \end{bmatrix} \tag{4-11}$$

The matrix in Eq. (4-11) is called the *orientation matrix* and is used to transform a vector between two consecutive links according to Eq. (4-10). This matrix describes the orientation of frame O_i with respect to O_{i-1} and is used to define the orientation of the link. It is also called the *rotation matrix* or the *direction cosine matrix* (DCM).

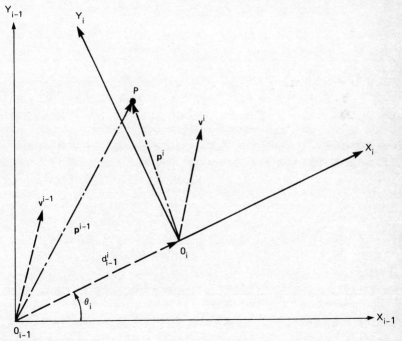

FIG. 4-3 Vector and point transformations between two planar links' coordinate systems.

The inverse transformation of Eq. (4-10) is used to transform the vector components from frame O_{i-1} to O_i according to

$$\mathbf{v}^i = \mathbf{C}_i^{i-1}\mathbf{v}^{i-1} \tag{4-12}$$

where $\mathbf{C}_i^{i-1}$ is the orientation matrix of O_{i-1} with respect to the O_i frame given by

$$\mathbf{C}_i^{i-1} = [\mathbf{C}_{i-1}^i]^{-1} = [\mathbf{C}_{i-1}^i]^{\mathrm{T}} \tag{4-13}$$

where $^{-1}$ denotes the inverse and ^{T} the transpose of a matrix. The inverse of the orientation matrix is equal to its transpose since the orientation matrices are orthogonal. The inverse transformation of Eq. (4-9), according to Eq. (4-13), is

$$\begin{bmatrix} v_x^i \\ v_y^i \end{bmatrix} = \begin{bmatrix} \cos\theta_i & \sin\theta_i \\ -\sin\theta_i & \cos\theta_i \end{bmatrix} \begin{bmatrix} v_x^{i-1} \\ v_y^{i-1} \end{bmatrix} \tag{4-14}$$

Note that if joint i were a prismatic joint, the orientation matrix between link i and link $i - 1$ is the identity matrix, which indicates no change in the orientation.

Chaining Orientation Matrices.   A vector $\mathbf{v}^3$ in frame O_3 in Fig. 4-2a can be expressed in O_2 by using Eq. (4-9) for $i = 3$, or in the form of Eq. (4-10)

$$\mathbf{v}^2 = \mathbf{C}_2^3\mathbf{v}^3 \tag{4-15}$$

Likewise $\mathbf{v}^2$ can be transformed to O_1 frame according to

$$\mathbf{v}^1 = \mathbf{C}_1^2\mathbf{v}^2 \tag{4-16}$$

where the matrices $\mathbf{C}$ are given in Eq. (4-11). Equations (4-15) and (4-16) can be combined to

$$\mathbf{v}^1 = \mathbf{C}_1^2\mathbf{C}_2^3\mathbf{v}^3 \tag{4-17}$$

or

$$\mathbf{v}^1 = \mathbf{C}_1^3\mathbf{v}^3 \tag{4-18}$$

The matrix $\mathbf{C}_1^3$ is the rotation transformation matrix from frame O_3 to frame O_1. The proof is based on matrix multiplication:

$$\mathbf{C}_1^3 = \mathbf{C}_1^2\mathbf{C}_2^3$$

$$= \begin{bmatrix} \cos\theta_2 & -\sin\theta_2 \\ \sin\theta_2 & \cos\theta_2 \end{bmatrix} \begin{bmatrix} \cos\theta_3 & -\sin\theta_3 \\ \sin\theta_3 & \cos\theta_3 \end{bmatrix}$$

$$= \begin{bmatrix} \cos\theta_2\cos\theta_3 - \sin\theta_2\sin\theta_3 & -\sin\theta_2\cos\theta_3 - \cos\theta_2\sin\theta_3 \\ \sin\theta_2\cos\theta_3 + \cos\theta_2\sin\theta_3 & \cos\theta_2\cos\theta_3 - \sin\theta_2\sin\theta_3 \end{bmatrix}$$

$$= \begin{bmatrix} \cos \psi & -\sin \psi \\ \sin \psi & \cos \psi \end{bmatrix} \tag{4-19}$$

where $\psi = \theta_2 + \theta_3$. By observing Fig. 4-2$a$, it can be seen that the orientation matrix $\mathbf{C}_1^3$ in Eq. (4-19) describes the orientation of O_3 with respect to the O_1 frame.

The operation described in Eqs. (4-17) and (4-19) is called a *chaining operation*, the result of which is an orientation matrix between two links not sharing the same joint. The orientation matrix can be defined between any two coordinate systems O_i and O_j, $\forall$ $i > j$ by using the following chaining operation

$$\mathbf{C}_j^i = \mathbf{C}_j^{j+1} \mathbf{C}_{j+1}^{j+2} \cdot \cdot \cdot \mathbf{C}_{i-2}^{i-1} \mathbf{C}_{i-1}^i \tag{4-20}$$

The resultant matrix $\mathbf{C}_j^i$ describes the orientation of frame O_i (or link i) with respect to O_j (or link j). In general, the transformation can be written in the form of Eq. (4-10)

$$\mathbf{v}^j = \mathbf{C}_j^i \mathbf{v}^i \tag{4-21}$$

which transforms a vector from O_i to O_j frame. A similar procedure is applied in the three-dimensional case, where $\mathbf{C}_j^i$ is a 3×3 matrix.

The Translation Vector and the Position Vector. The translation vector $\mathbf{d}_{i-1}^i$ (see Fig. 4-3) describes the position of the origin of frame O_i in O_{i-1} and is given by

$$\mathbf{d}_{i-1}^i = \begin{bmatrix} a_i \cos \theta_i \\ a_i \sin \theta_i \end{bmatrix} \tag{4-22}$$

where a_i is the distance between O_i and O_{i-1}.

A point P which is defined in O_i by the position vector $\mathbf{p}^i$ can be expressed in O_{i-1} by the vector

$$\mathbf{p}^{i-1} = \mathbf{d}_{i-1}^i + \mathbf{C}_{i-1}^i \mathbf{p}^i \tag{4-23}$$

Note that $\mathbf{p}^i$ has been first expressed in O_{i-1} using Eq. (4-10), and then a vector summation has been performed in the O_{i-1} frame.

4.3.2 Homogeneous Transformation Matrices

The homogeneous representation of a two-dimensional vector includes three components

$$\mathbf{v} = \begin{bmatrix} v_x \\ v_y \\ 1 \end{bmatrix} \tag{4-24}$$

The symbols of vectors from here through Sec. 4.5 are related to the homogeneous representation.

The homogeneous representations of two-dimensional transformation matrices are in the form of 3×3 matrices. Three types of matrices are used:

1. $T_R{}_{i-1}^i$ = the homogeneous *rotation* matrix which represents the rotation of frame O_i with respect to O_{i-1} and is used to transform a vector v from O_i to O_{i-1}.

2. $T_T{}_{i-1}^i$ = the homogeneous *translation* matrix, which represents the translation (no rotation) of frame O_i with respect to O_{i-1} and is used to transform a position vector **p** from O_i to O_{i-1}.

3. T_{i-1}^i = the homogeneous *displacement* matrix which represents the translation and rotation of frame O_i with respect to O_{i-1} and is used to transform a position vector **p** from O_i to O_{i-1}. The matrix **T** is also denoted as the Denavit-Hartenberg (DH) matrix.

The operation of the last matrix is similar to the combined operation of the previous two matrices. That means that a position vector can be either first rotated by using T_R and subsequently translated using T_T, or it can be directly transformed using the displacement matrix **T**.

The Homogeneous Rotation Matrix. The two-dimensional homogeneous rotation matrix between frame O_i (which is rotated with respect to O_{i-1}) and the O_{i-1} frame is

$$\mathbf{T_R}{}_{i-1}^i = \left[\begin{array}{cc|c} \mathbf{C}_{i-1}^i & & 0 \\ & & 0 \\ \hline 0 & 0 & 1 \end{array} \right] \tag{4-25}$$

where $\mathbf{C}_{i-1}^i$ is the 2×2 orientation matrix given in Eq. (4-11). The vector $\mathbf{v}^i$ (see Fig. 4-3) is transformed to $\mathbf{v}^{i-1}$ with the homogeneous rotation matrix as follows:

$$\begin{bmatrix} v_x^{i-1} \\ v_y^{i-1} \\ 1 \end{bmatrix} = \begin{bmatrix} \cos\theta_i & -\sin\theta_i & 0 \\ \sin\theta_i & \cos\theta_i & 0 \\ 0 & 0 & 1 \end{bmatrix} \begin{bmatrix} v_x^i \\ v_y^i \\ 1 \end{bmatrix} \tag{4-26}$$

or similar to Eq. (4-10):

$$\mathbf{v}^{i-1} = \mathbf{T_R}{}_{i-1}^i \mathbf{v}^i \tag{4-27}$$

The Homogeneous Translation Matrix. The 3×3 two-dimensional homogeneous translation matrix between O_i (which is translated parallel to O_{i-1}) and the O_{i-1} frame is

$$\mathbf{T}_{T}{}^{i}_{i-1} = \begin{bmatrix} 1 & 0 & | & \mathbf{d}^i_{i-1} \\ 0 & 1 & | & \\ \hline 0 & 0 & | & 1 \end{bmatrix} \qquad (4\text{-}28)$$

where $\mathbf{d}^i_{i-1}$ is the translation vector given in Eq. (4-22).

As an example assume that frame O_i is translated in the amount a_i parallel to O_{i-1} as shown in Fig. 4-4. The origin of O_i is described in O_{i-1} by the translation $\mathbf{d}^i_{i-1}$ given in Eq. (4-22). A point P is described in O_i and O_{i-1} by the position vectors $\mathbf{p}^i$ and $\mathbf{p}^{i-1}$, respectively. The vector $\mathbf{p}^i$ is transformed to $\mathbf{p}^{i-1}$ with the homogeneous translation matrix in Eq. (4-28) as follows:

$$\begin{bmatrix} p_x^{i-1} \\ p_y^{i-1} \\ 1 \end{bmatrix} = \begin{bmatrix} 1 & 0 & a_i \cos \theta_i \\ 0 & 1 & a_i \sin \theta_i \\ 0 & 0 & 1 \end{bmatrix} \begin{bmatrix} p_x^i \\ p_y^i \\ 1 \end{bmatrix} = \begin{bmatrix} p_x^i + a_i \cos \theta_i \\ p_y^i + a_i \sin \theta_i \\ 1 \end{bmatrix} \qquad (4\text{-}29)$$

Note that this transformation performs a vector summation of $\mathbf{d}^i_{i-1}$ and $\mathbf{p}^i$. The ability to perform vector summation in a matrix form is a consequence of the component 1 which is added in the homogeneous representation. This feature differs the homogeneous transformations from other transformation methods.

The Homogeneous Displacement Matrix. The homogeneous displacement matrix $\mathbf{T}^i_{i-1}$ transforms a point coordinate expressed in O_i to O_{i-1}. Referring to Fig. 4-3, the position vector $\mathbf{p}^{i-1}$ can be expressed in

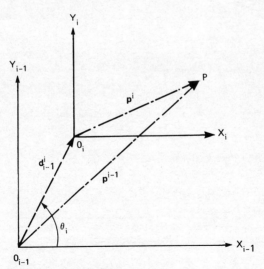

FIG. 4-4 Coordinate transformation between planar coordinate systems that are translated relative to each other.

terms of $\mathbf{p}^i$ by first rotating $\mathbf{p}^i$, thus expressing it in a coordinate system parallel to the O_{i-1} frame, and then translating the rotated $\mathbf{p}^i$ by summing it with $\mathbf{d}^i_{i-1}$. Using homogeneous transformation matrices, this operation is performed as follows:

$$\mathbf{p}^{i-1} = \mathbf{T}_{T\,i-1}^{\quad i} \mathbf{T}_{R\,i-1}^{\quad i} \mathbf{p}^i \tag{4-30}$$

or

$$\mathbf{p}^{i-1} = \mathbf{T}^i_{i-1}\mathbf{p}^i \tag{4-31}$$

with $\mathbf{T}^i_{i-1}$ being the homogeneous displacement matrix between O_i and O_{i-1} derived from

$$\mathbf{T}^i_{i-1} = \mathbf{T}_{T\,i-1}^{\quad i} \mathbf{T}_{R\,i-1}^{\quad i} \tag{4-32}$$

Note that Eq. (4-30) performs the same transformation as Eq. (4-23) involving only matrix operations. Multiplying the matrices in Eq. (4-32) for $\mathbf{T}_T$ and $\mathbf{T}_R$ given in Eqs. (4-28) and (4-25), respectively, yields

$$\mathbf{T}^i_{i-1} = \left[\begin{array}{c|c} \mathbf{C}^i_{i-1} & \mathbf{d}^i_{i-1} \\ \hline 0 \quad 0 & 1 \end{array} \right] \tag{4-33}$$

which is the standard form of the displacement matrix. The matrix $\mathbf{C}^i_{i-1}$ and the vector $\mathbf{d}^i_{i-1}$ describe the orientation and the position of O_i in respect to O_{i-1}, respectively.

For the two coordinate systems in Fig. 4-3, the matrix $\mathbf{C}^i_{i-1}$ in Eq. (4-11) and the vector $\mathbf{d}^i_{i-1}$ in Eq. (4-22) are substituted in Eq. (4-33) yielding

$$\mathbf{T}^i_{i-1} = \left[\begin{array}{ccc} \cos\theta_i & -\sin\theta_i & a_i\cos\theta_i \\ \sin\theta_i & \cos\theta_i & a_i\sin\theta_i \\ 0 & 0 & 1 \end{array} \right] \tag{4-34}$$

For any two coordinate systems O_i and O_j the displacement matrix is

$$\mathbf{T}^i_j = \left[\begin{array}{c|c} \mathbf{C}^i_j & \mathbf{d}^i_j \\ \hline 0 \quad 0 & 1 \end{array} \right] \tag{4-35}$$

where $\mathbf{C}^i_j$ describes the orientation of O_i in frame O_j, and $\mathbf{d}^i_j$ is the position vector of the origin of O_i in O_j.

The transformation of a position vector $\mathbf{p}^i$ in O_i to $\mathbf{p}^j$ in the O_j frame is performed according to

$$\mathbf{p}^j = \mathbf{T}^i_j\mathbf{p}^i \tag{4-36}$$

and

$$\mathbf{T}^i_j = \mathbf{T}_{T\,j}^{\quad i} \mathbf{T}_{R\,j}^{\quad i} \tag{4-37}$$

Inverse Homogeneous Transformation.   The inverse transformation of that shown in Eq. (4-36) is performed according to

$$\mathbf{p}^i = \mathbf{T}^j_i \mathbf{p}^j \tag{4-38}$$

Since

$$\mathbf{T}^j_i = [\mathbf{T}^i_j]^{-1} \tag{4-39}$$

an inverse of the displacement matrix is required. To facilitate the inversion of $\mathbf{T}^i_j$, the translation and rotation matrices can be applied.

Implementing the properties of matrix inversion,

$$[\mathbf{A}\quad\mathbf{B}]^{-1} = \mathbf{B}^{-1}\mathbf{A}^{-1} \tag{4-40}$$

on Eqs. (4-37) yields

$$[\mathbf{T}^i_j]^{-1} = [\mathbf{T}_R{}^i_j]^{-1}[\mathbf{T}_T{}^i_j]^{-1} \tag{4-41}$$

It can be shown that the inverse of the homogeneous rotation and translation matrices are

$$[\mathbf{T}_R{}^i_j]^{-1} = \left[\begin{array}{ccc|c} & [\mathbf{C}^i_j]^T & & 0 \\ & & & 0 \\ \hline 0 & 0 & & 1 \end{array}\right] \tag{4-42}$$

and

$$[\mathbf{T}_T{}^i_j]^{-1} = \left[\begin{array}{cc|c} 1 & 0 & -\mathbf{d}^i_j \\ 0 & 1 & \\ \hline 0 & 0 & 1 \end{array}\right] \tag{4-43}$$

Referring to Fig. 4-3, the inverse transformations presented in Eqs. (4-42) and (4-43) are derived (for $j = i - 1$)

$$[\mathbf{T}_R{}^i_{i-1}]^{-1} = \left[\begin{array}{ccc} \cos\theta_i & \sin\theta_i & 0 \\ -\sin\theta_i & \cos\theta_i & 0 \\ 0 & 0 & 1 \end{array}\right] \tag{4-44}$$

$$[\mathbf{T}_T{}^i_{i-1}]^{-1} = \left[\begin{array}{ccc} 1 & 0 & -a_i\cos\theta_i \\ 0 & 1 & -a_i\sin\theta_i \\ 0 & 0 & 1 \end{array}\right] \tag{4-45}$$

Substituting Eqs. (4-44) and (4-45) into Eq. (4-41) for $j = i - 1$ yields

$$[\mathbf{T}^i_{i-1}]^{-1} = \left[\begin{array}{ccc} \cos\theta_i & \sin\theta_i & -a_i \\ -\sin\theta_i & \cos\theta_i & 0 \\ 0 & 0 & 1 \end{array}\right] \tag{4-46}$$

By definition, from Eq. (4-35)

$$\mathbf{T}_i^{i-1} = \left[\begin{array}{c|c} \mathbf{C}_i^{i-1} & \mathbf{d}_i^{i-1} \\ \hline 0 \quad 0 & 1 \end{array} \right] \tag{4-47}$$

Equations (4-46) and (4-47) are identical since

$$\mathbf{d}_i^{i-1} = \left[\begin{array}{c} -a_i \\ 0 \end{array} \right] \quad \text{and} \quad \mathbf{C}_i^{i-1} = [\mathbf{C}_{i-1}^i]^\mathrm{T} \tag{4-48}$$

for $\mathbf{d}_i^{i-1}$ describing the position of the origin of O_{i-1} in O_i frame, and $\mathbf{C}_i^{i-1}$ describes the orientation of O_{i-1} in O_i frame [see Eq. (4-13)].

Chaining Displacement Matrices.   The chaining operation of homogeneous displacement matrices is similar to chaining of orientation matrices. For two transformations

$$\mathbf{p}^j = \mathbf{T}_j^i \mathbf{p}^i \tag{4-49}$$

and

$$\mathbf{p}^i = \mathbf{T}_i^k \mathbf{p}^k \tag{4-50}$$

the chaining operation is performed by substituting $\mathbf{p}^i$ from Eq. (4-50) to Eq. (4-49) which yields

$$\mathbf{p}^j = \mathbf{T}_j^i \mathbf{T}_i^k \mathbf{p}^k \tag{4-51}$$

where

$$\mathbf{T}_j^k = \mathbf{T}_j^i \mathbf{T}_i^k \tag{4-52}$$

or in general form, for $i > j$

$$\mathbf{T}_j^i = \mathbf{T}_j^{j+1} \mathbf{T}_{j+1}^{j+2} \cdots \mathbf{T}_{i-2}^{i-1} \mathbf{T}_{i-1}^i \tag{4-53}$$

Referring to the three coordinate systems in Fig. 4-2a, the two displacement matrices $\mathbf{T}_1^2$ and $\mathbf{T}_2^3$ are in the form of Eq. (4-34) for $i = 2$ and $i = 3$, respectively. Transforming the position vector of point P from O_3 to O_1 using the chaining operation according to Eq. (4-51) yields

$$\mathbf{p}^1 = \mathbf{T}_1^3 \mathbf{p}^3 \tag{4-54}$$

where

$$\mathbf{T}_1^3 = \mathbf{T}_1^2 \mathbf{T}_2^3 = \left[\begin{array}{cc|c} \cos\psi & -\sin\psi & a_3\cos\psi + a_2\cos\theta_2 \\ \sin\psi & \cos\psi & a_3\sin\psi + a_2\sin\theta_2 \\ \hline 0 & 0 & 1 \end{array} \right] \tag{4-55}$$

and $\psi = \theta_2 + \theta_3$ [see Eq. (4-19)].

The TCP is located at the origin of the O_3 frame and therefore its position relative to O_1 (the reference coordinate system) is represented by the translation vector

$$\begin{bmatrix} X_t \\ Y_t \end{bmatrix} = \mathbf{d}_1^3 = \begin{bmatrix} a_2 \cos \theta_2 & + & a_3 \cos \psi \\ a_2 \sin \theta_2 & + & a_3 \sin \psi \end{bmatrix} \tag{4-56}$$

which is derived from $\mathbf{T}_1^3$ in Eq. (4-55). Equation (4-56) is identical to Eq. (4-1) which was derived from simple geometrical considerations. The orientation matrix $\mathbf{C}_1^3$ is identical to Eq. (4-19), and it can be seen that it describes the orientation of link 3 with respect to the O_1 frame.

4.3.3 Three-Dimensional Homogeneous Transformations

The homogenous representation of a three-dimensional vector contains four components

$$\mathbf{v} = \begin{bmatrix} v_x \\ v_y \\ v_z \\ 1 \end{bmatrix} \tag{4-57}$$

The homogeneous representation of three-dimensional transformation matrices includes the homogeneous rotation, translation, and displacement matrices, which are in the form of 4×4 matrices.

Rotation Matrix. The rotation transformation matrix between two coordinate systems O_i and O_{i-1} is

$$\mathbf{T}_{R\ i-1}^{\ i} = \left[\begin{array}{ccc|c} & & & 0 \\ & \mathbf{C}_{i-1}^i & & 0 \\ & & & 0 \\ \hline 0 & 0 & 0 & 1 \end{array} \right] \tag{4-58}$$

where O_i rotates relative to frame O_{i-1}.

There are three rotation matrices for describing the rotation of O_i about the X_{i-1}, Y_{i-1}, and Z_{i-1} axes. For rotating O_i through an angle α_i about the X_{i-1} axis, the following transformation matrix is used:

$$[\mathbf{T}_{R\ i-1}^{\ i}]_x = \left[\begin{array}{ccc|c} 1 & 0 & 0 & 0 \\ 0 & \cos \alpha_i & -\sin \alpha_i & 0 \\ 0 & \sin \alpha_i & \cos \alpha_i & 0 \\ \hline 0 & 0 & 0 & 1 \end{array} \right] \tag{4-59}$$

Likewise, for rotating O_i through an angle β_i about the Y_{i-1} axis

$$[\mathbf{T}_{R\ i-1}^{\ i}]_y = \left[\begin{array}{ccc|c} \cos \beta_i & 0 & \sin \beta_i & 0 \\ 0 & 1 & 0 & 0 \\ -\sin \beta_i & 0 & \cos \beta_i & 0 \\ \hline 0 & 0 & 0 & 1 \end{array} \right] \tag{4-60}$$

and for rotating O_i through an angle θ_i about the Z_{i-1} axis

$$[\mathbf{T}_{R\,i-1}^{\ i}]_z = \left[\begin{array}{ccc|c} \cos\theta_i & -\sin\theta_i & 0 & 0 \\ \sin\theta_i & \cos\theta_i & 0 & 0 \\ 0 & 0 & 1 & 0 \\ \hline 0 & 0 & 0 & 1 \end{array}\right] \tag{4-61}$$

Translation Matrix. The homogeneous translation matrix for the three-dimensional case is

$$\mathbf{T}_{T\,i-1}^{\ i} = \left[\begin{array}{ccc|c} 1 & 0 & 0 & \\ 0 & 1 & 0 & \mathbf{d}_{i-1}^i \\ 0 & 0 & 1 & \\ \hline 0 & 0 & 0 & 1 \end{array}\right] \tag{4-62}$$

where $\mathbf{d}_{i-1}^i$ describes the origin of O_i in frame O_{i-1}. Equation (4-62) is equivalent to Eq. (4-28) of the planar case.

Displacement Matrix. The homogeneous displacement matrix is obtained by successive rotation and translation of O_i in respect to frame O_{i-1}:

$$\mathbf{T}_{i-1}^i = \mathbf{T}_{T\,i-1}^{\ i}\,\mathbf{T}_{R\,i-1}^{\ i} \tag{4-63}$$

Substituting Eqs. (4-58) and (4-62) into (4-63) yields:

$$\mathbf{T}_{i-1}^i = \left[\begin{array}{ccc|c} & \mathbf{C}_{i-1}^i & & \mathbf{d}_{i-1}^i \\ \hline 0 & 0 & 0 & 1 \end{array}\right] \tag{4-64}$$

For any two coordinate systems O_i and O_j the displacement matrix is $\mathbf{T}_j^i$ where $j = i - 1$ in Eq. (4-64).

Vector Transformation. The vector transformation between two coordinate systems O_i and O_{i-1} is similar to that in Eq. (4-27) for the planar case

$$\mathbf{v}^{i-1} = \mathbf{T}_{R\,i-1}^{\ i}\,\mathbf{v}^i \tag{4-65}$$

where $\mathbf{v}^i$ (expressed in O_i) is transformed to $\mathbf{v}^{i-1}$ (expressed in O_{i-1}), and $\mathbf{T}_{R\,i-1}^{\ i}$ is given in Eq. (4-58).

As an example, consider the three coordinate systems O_{i-1}, O_i' and O_i described in Fig. 4-5. Frame O_i' is translated relative to O_{i-1} but remains parallel to it. Frame O_i is obtained by a rotation around X_i' at an angle α_i. The vector v can be described by $\mathbf{v}^i$ or $\mathbf{v}^{i-1}$. The rotation transformation matrix between O_i and O_{i-1} is given in Eq. (4-59) and is used to transform $\mathbf{v}^i$ to $\mathbf{v}^{i-1}$ according to Eq. (4-65).

$$\begin{bmatrix} v_x^{i-1} \\ v_y^{i-1} \\ v_z^{i-1} \\ 1 \end{bmatrix} = \begin{bmatrix} 1 & 0 & 0 & 0 \\ 0 & \cos\alpha_i & -\sin\alpha_i & 0 \\ 0 & \sin\alpha_i & \cos\alpha_i & 0 \\ 0 & 0 & 0 & 1 \end{bmatrix} \begin{bmatrix} v_x^i \\ v_y^i \\ v_z^i \\ 1 \end{bmatrix} \qquad (4\text{-}66)$$

For $\alpha_i = \pi/2$ Eq. (4-66) becomes

$$\begin{bmatrix} v_x^{i-1} \\ v_y^{i-1} \\ v_z^{i-1} \\ 1 \end{bmatrix} = \begin{bmatrix} 1 & 0 & 0 & 0 \\ 0 & 0 & -1 & 0 \\ 0 & 1 & 0 & 0 \\ 0 & 0 & 0 & 1 \end{bmatrix} \begin{bmatrix} v_x^i \\ v_y^i \\ v_z^i \\ 1 \end{bmatrix} = \begin{bmatrix} v_x^i \\ -v_z^i \\ v_y^i \\ 1 \end{bmatrix} \qquad (4\text{-}67)$$

which means that in this case the projection of **v** on Y_i (i.e., v_y^i) is seen in frame O_{i-1} on axis Z_{i-1}, and v_z^i is seen in O_{i-1} on the $-Y_{i-1}$ axis.

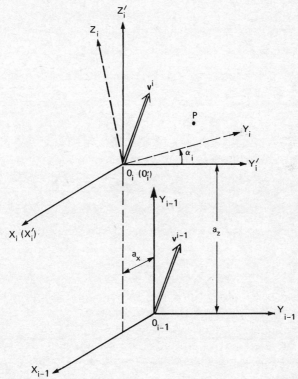

FIG. 4-5 Vector and coordinate transformations between two links' coordinate systems that are translated relative to each other.

Point Coordinates Transformation   A point is described by a position vector **p** and is transformed from O_i to O_{i-1} using the displacement transformation matrix.

$$\mathbf{p}^{i-1} = \mathbf{T}^i_{i-1}\mathbf{p}^i \tag{4-68}$$

where $\mathbf{T}^i_{i-1}$ is given in Eq. (4-64). Equation (4-68) is equivalent to Eq. (4-31) of the planar case. The transformation given in Eq. (4-68) is also called *coordinate transformation*.

For example, assume that the point P in Fig. 4-5 is given in frame O_i and should be expressed in frame O_{i-1}. The transformation is performed in two steps according to Eq. (4-63). First $\mathbf{p}^i$ is rotated to frame O_i^1 using the rotation matrix in Eq. (4-59) and then the obtained $\mathbf{p}^{i'}$ is translated using the translation matrix. The obtained transformation matrix is calculated as follows:

$$\mathbf{T}^i_{i-1} = \begin{bmatrix} 1 & 0 & 0 & a_x \\ 0 & 1 & 0 & 0 \\ 0 & 0 & 1 & a_z \\ 0 & 0 & 0 & 1 \end{bmatrix} \begin{bmatrix} 1 & 0 & 0 & 0 \\ 0 & \cos\alpha_i & -\sin\alpha_i & 0 \\ 0 & \sin\alpha_i & \cos\alpha_i & 0 \\ 0 & 0 & 0 & 1 \end{bmatrix} \tag{4-69}$$

and consequently Eq. (4-68) for this example is

$$\mathbf{p}^{i-1} = \begin{bmatrix} 1 & 0 & 0 & a_x \\ 0 & \cos\alpha_i & -\sin\alpha_i & 0 \\ 0 & \sin\alpha_i & \cos\alpha_i & a_z \\ 0 & 0 & 0 & 1 \end{bmatrix} \mathbf{p}^i \tag{4-70}$$

The inverse homogeneous transformations and the chaining operation presented for the planar case are also valid for three-dimensional mechanisms. The chaining is performed by multiplying the associated matrices according to Eq. (4-53) obtaining

$$\mathbf{T}^i_j = \left[\begin{array}{ccc|c} & C^i_j & & d^i_j \\ \hline 0 & 0 & 0 & 1 \end{array} \right] \tag{4-71}$$

Summary.   The homogeneous transformation between each two consecutive coordinate systems (O_i and O_{i-1}) is presented in the form of a matrix $\mathbf{T}^i_{i-1}$ which is called the homogeneous displacement matrix or just the displacement matrix. The determination of the displacement matrix for two successive coordinate systems (O_{i-1} and O_i) can be obtained from geometric considerations as shown in Fig. 4-3.

The matrix $\mathbf{T}^i_{i-1}$ is composed of a translation vector $\mathbf{d}^i_{i-1}$ and an orientation matrix $\mathbf{C}^i_{i-1}$, both having geometric interpretation. The components of the translation vector $\mathbf{d}^i_{i-1}$ express the projections of the origin of frame O_i on O_{i-1}. The orientation matrix (sometimes called the rotation matrix) $\mathbf{C}^i_{i-1}$ is used to describe the orientation of O_i with respect to the O_{i-1} frame. It can also be used to transform vectors between the two coordinate systems. A vector v can be represented as a directed line segment and is expressed either as $\mathbf{v}^{i-1}$ in frame O_{i-1} or as $\mathbf{v}^i$ in O_i. In both representations the vector has the same magnitude and the same absolute direction. The orientation matrix is used to transform the vector $\mathbf{v}^i$ to $\mathbf{v}^{i-1}$. The displacement matrix $\mathbf{T}_{i-1}^i$ is used for the following:

1. Transforming point coordinates from frame O_i to O_j using Eq. (4-36) or Eq. (4-68), or from O_j to O_i using Eq. (4-38). This feature is used, for example, when the position of an object is detected by a sensor placed in the end effector and it is necessary to express its position in the WCS.

2. Expressing in the O_j frame a vector v given in the O_i frame, using the rotation transformation in Eq. (4-27) or Eq. (4-65). It can be used, for example, when a desired velocity of the TCP is defined in the TCS and has to be transformed to the WCS. This situation occurs when a sensor is measuring in TCS, or when teaching the robot in TCS.

3. Expressing the position of the origin of O_i in O_j, which is represented by $\mathbf{d}_j^i$.

4. Defining the orientation of the link to which O_i frame is attached relative to O_j frame. The definition is given by the matrix $\mathbf{C}_j^i$.

Two alternative methods to solve the arm kinematics have been presented. The straightforward geometric technique is simpler for mechanisms containing a small number of axes in which only the position of the TCP is of concern. The mathematical equations for this technique are developed separately for each particular mechanism. By contrast, the homogeneous transformation is a general method in which the complexity of the solution is independent of the number of axes. The algorithm consists mainly of matrix multiplication and therefore is easily implemented. The solution contains, in addition to the position of the TCP, also the orientation of the tool and the position of any point on the links, a feature which is especially important when the arm dynamics is of concern. Therefore, this method is sometimes useful even for mechanisms containing a small number of axes.

4.4 DENAVIT-HARTENBERG CONVENTION

The homogeneous displacement matrix $\mathbf{T}_{i-1}^i$ presented in Sec. 4.3 should be derived for each particular case using Eqs. (4-59) to (4-62).

A systematic technique for establishing the displacement matrix for each two adjacent links of a mechanism was proposed by Denavit and Hartenberg (DH) in 1955 (Denavit and Hartenberg, 1955).

4.4.1 Implementing the DH Convention

The DH convention is mainly implemented in robot manipulators which consist of an open kinematic chain in which each joint contains one degree of freedom and the joint is either revolute or prismatic.

The revolute and prismatic joints are considered as *lower pairs*, i.e., joints which have two surfaces sliding over one another while remaining in contact. The six possible lower-pair types are revolute, prismatic, cylindrical, spherical (ball and socket), screw, and planar pair, of which only the revolute and prismatic are typically used in robot manipulators.

The DH convention is implemented through the following steps:

1. *Number the links and joints*, starting at the base. The stationary base is denoted as link O and the end effector is link n, as demonstrated in Fig. 4-2b. Link i moves in respect to link $i - 1$ around (for revolute) or along (for prismatic) joint i (see Fig. 4-6).

2. *Establish links' coordinate systems* for each of the joints according to the following rules (see Figs. 4-2 and 4-6):

   a. The Z_{i-1} axis is chosen along the axis of motion of joint i. For a revolute joint, link i rotates in respect to link $i - 1$ around the $+Z_{i-1}$ axis in the amount $+\theta_i$; for a prismatic joint, link i is displaced relative to link $i - 1$ along the $+Z_{i-1}$ axis in the amount $+d_i$.

   b. The X_i axis is chosen perpendicular to the Z_{i-1} axis (i.e., it is perpendicular to both Z_{i-1} and Z_i). If Z_i and Z_{i-1} do not intersect, then the X_i axis is along the common normal to Z_i and Z_{i-1} and its direction is defined from Z_{i-1} toward the Z_i axis. If, however, Z_{i-1} and Z_i do intersect, the direction of X_i axis is not defined and it can be chosen in either of the two possible directions. In addition, if the Z_{i-1} and Z_i axes are colinear, the X_i axis can be chosen anywhere in the plane perpendicular to them.

   c. The Y_i axis is chosen to complete a right-handed coordinate system.

Note that the assignment of coordinate systems is not unique. For example, there are several possibilities for the selection of the direction of the X_i axis.

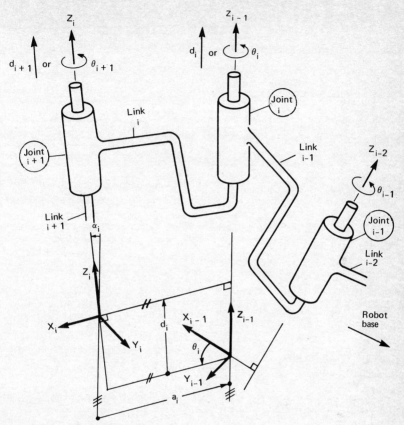

FIG. 4-6 Links' coordinate systems and joint parameters according to Denavit and Hartenberg method.

3. *Define the joint parameters* which are the four geometric quantities θ_i, d_i, a_i, α_i (see Fig. 4-6).

θ_i is the angle between the X_{i-1} and the X_i axis, obtained by screwing X_{i-1} into X_i around the Z_{i-1} axis. For a revolute joint, θ_i is a variable, and for a prismatic joint, θ_i is a constant parameter.

d_i is the coordinate of the origin of O_i frame on the Z_{i-1} axis, i.e., the distance between the origin of O_{i-1} frame to the intersection of the Z_{i-1} axis with the X_i axis. For a prismatic joint d_i is a variable, and for a revolute joint d_i is a constant parameter.

a_i is the distance between Z_{i-1} and Z_i axis measured along the negative direction of X_i from its origin to where it intersects the Z_{i-1} axis (a constant parameter).

α_i is the angle between the Z_{i-1} axis and the Z_i axis, obtained by screwing Z_{i-1} into Z_i around the X_i axis (a constant parameter).

Note that there are link configurations for which it is impossible to establish the joint parameters according to the above definitions,[†] and another method should be applied.

4. *Form the homogeneous displacement matrix* for joint i by applying Eq. (4-64)

$$\mathbf{T}_{i-1}^i = \left[\begin{array}{cc|c} & \mathbf{C}_{i-1}^i & \mathbf{d}_{i-1}^i \\ \hline 0 & 0 & 0 & 1 \end{array} \right] \tag{4-72}$$

The upper left 3×3 portion of $\mathbf{T}_{i-1}^i$ is the orientation matrix $\mathbf{C}_{i-1}^i$ of link i with respect to link $i - 1$ and is given by

$$\mathbf{C}_{i-1}^i = \left[\begin{array}{ccc} \cos\theta_i & -\cos\alpha_i \sin\theta_i & \sin\alpha_i \sin\theta_i \\ \sin\theta_i & \cos\alpha_i \cos\theta_i & -\sin\alpha_i \cos\theta_i \\ 0 & \sin\alpha_i & \cos\alpha_i \end{array} \right] \tag{4-73}$$

The orientation matrix is also denoted as the DCM of link i with respect to link $i - 1$. In principle, three independent parameters are required to define the orientation between any two coordinate systems. However, since $\mathbf{C}_{i-1}^i$ in Eq. (4-73) contains only two independent parameters (θ_i and α_i), it can be used only for coordinate systems that are oriented to each other by two consecutive rotations θ_i (first) and then α_i. Note that with lower-pair joints these two rotations are sufficient. If the links' coordinate systems are assigned according to the DH convention, it is always possible to derive the orientation matrix according to Eq. (4-73).

The three upper components on the right column of $\mathbf{T}_{i-1}^i$ in Eq. (4-72) are the components of the translation vector $\mathbf{d}_{i-1}^i$

$$\mathbf{d}_{i-1}^i = \left[\begin{array}{c} a_i \cos\theta_i \\ a_i \sin\theta_i \\ d_i \end{array} \right] \tag{4-74}$$

The vector $\mathbf{d}_{i-1}^i$ describes the position of the origin of frame O_i expressed in frame O_{i-1}. In robotics there are cases in which $\mathbf{d}_{i-1}^i$ cannot be derived according to convention. In such cases an alternative technique is used as will be shown in the following sections.

[†]Denavit and Hartenberg presented an additional rule which dictates the location of the origin of the coordinate systems. This enables the definition of joint parameters for any lower-pair joint configuration. In order to facilitate the understanding and implementation of the DH technique, this rule was omitted in our text.

In any case, d_{i-1}^i always describes the location of the origin of frame O_i in O_{i-1}.

The implementation of Eqs. (4-72) to (4-74) will be shown in the following section.

4.4.2 Obtaining the DH Displacement Matrices

The displacement matrix in Eqs. (4-72) to (4-74) can be obtained by applying successive rotations and translations to frame O_{i-1} to align it with O_i. This requires the following sequence of operations:

1. Rotate O_{i-1} around the Z_{i-1} axis at an angle θ_i.
2. Translate the rotated frame O_{i-1} by a_i and d_i units in the directions X_{i-1} and Z_{i-1}, respectively.
3. Rotate frame O_{i-1} (formed in step 2) at an angle α_i around X_{i-1} to form the frame O_i.

Applying the transformation matrices in Eqs. (4-61), (4-62), and (4-59) successively yields

$$
T_{i-1}^i =
\begin{bmatrix}
\cos\theta_1 & -\sin\theta_1 & 0 & 0 \\
\sin\theta_1 & \cos\theta_1 & 0 & 0 \\
0 & 0 & 1 & 0 \\
0 & 0 & 0 & 1
\end{bmatrix}
$$

$$
\cdot
\begin{bmatrix}
1 & 0 & 0 & a_i \\
0 & 1 & 0 & 0 \\
0 & 0 & 1 & d_i \\
0 & 0 & 0 & 1
\end{bmatrix}
\cdot
\begin{bmatrix}
1 & 0 & 0 & 0 \\
0 & \cos\alpha_i & -\sin\alpha_i & 0 \\
0 & \sin\alpha_i & \cos\alpha_i & 0 \\
0 & 0 & 0 & 1
\end{bmatrix}
\tag{4-75}
$$

Performing the matrix multiplication results in the DH displacement transformation matrix in Eqs. (4-72) to (4-74).

4.5 APPLICATIONS OF THE DH METHOD

This section presents examples which demonstrate the derivation of displacement matrices according to the DH convention presented in Sec. 4.4.

Robot manipulators consist of an arm and a wrist. The arm is commonly referred to the links moved by the first three joints (starting from the base), and the wrist is referred to the remaining links. Since various types of arms and wrists can be combined to form a manipulator, they are discussed separately. The transformation matrix of the manipulator is obtained by chaining the transformation matrices of the associated arm and wrist.

4.5.1 Three-Axis Robot Arms

The articulated and spherical arms are discussed in this section.

An Articulated Arm.   A three-axis articulated arm with three revolute joints is schematically shown in Fig. 4-7. In the illustrated position all joint variables are at their reference position, i.e., $\theta_i = 0$ for $i = 1$ to 3. The corresponding transformation matrices are derived according to the procedure defined in Sec. 4.4:

1.   The links are numbered, as shown in Fig. 4-7.

2.   Coordinate systems are assigned to the various links: The Z_i axes point to the joints' axes of rotation. The X_i axes are chosen perpendicular to both Z_{i-1} and Z_i axes. Coordinate system O_3, at the end of the

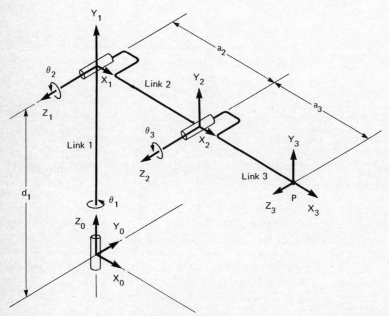

FIG. 4-7   Links' coordinate systems and joint parameters for articulated arm.

TABLE 4-1 Joint Parameters for the Articulated Arm

Joint	θ_i†	d_i	a_i	α_i
	Joint parameters			
1	θ_1	d_1	0	$+90°$
2	θ_2	0	a_2	0
3	θ_3	0	a_3	0

†Variables.

arm, corresponds to a wrist having its first joint rotating about, or sliding along, the Z_3 axis.

3. The joint parameters are established as follows: The distance between the origins of O_0 and O_1 along the Z_0 axis is d_1. The axes Z_0 and Z_1 intersect and therefore $a_1 = 0$. The orientation angle α_1, measured from the Z_0 axis to Z_1 around X_1 (according to the right-hand rule) is $\alpha_1 = +90°$. The angle θ_1 (measured from X_0 to X_1) is the variable parameter of joint 1. Similarly, the parameters of the other joints are determined and are given in Table 4-1.

4. The displacement transformation matrices are determined, substituting the joint parameters from Table 4-1 to Eq. (4-72):

$$\mathbf{T}_0^1 = \begin{bmatrix} C_1 & 0 & S_1 & 0 \\ S_1 & 0 & -C_1 & 0 \\ 0 & 1 & 0 & d_1 \\ 0 & 0 & 0 & 1 \end{bmatrix} \tag{4-76}$$

$$\mathbf{T}_1^2 = \begin{bmatrix} C_2 & -S_2 & 0 & a_2C_2 \\ S_2 & C_2 & 0 & a_2S_2 \\ 0 & 0 & 1 & 0 \\ 0 & 0 & 0 & 1 \end{bmatrix} \tag{4-77}$$

$$\mathbf{T}_2^3 = \begin{bmatrix} C_3 & -S_3 & 0 & a_3C_3 \\ S_3 & C_3 & 0 & a_3S_3 \\ 0 & 0 & 1 & 0 \\ 0 & 0 & 0 & 1 \end{bmatrix} \tag{4-78}$$

where $C_i = \cos \theta_i$ and $S_i = \sin \theta_i$. $\tag{4-79}$

Now, the displacement matrix between frame O_3 and the base coordinate system (O_0) is established utilizing the chaining operation of Eq. (4-53) to the matrices in Eqs. (4-76) to (4-78), which yields

$$
\mathbf{T}_0^3 = \mathbf{T}_0^1\mathbf{T}_1^2\mathbf{T}_2^3 =
\begin{bmatrix}
\begin{array}{c}C_1C_2C_3 \\ +S_1C_2S_3\end{array} & \begin{array}{c}-C_1S_2S_3 \\ -S_1S_2C_3\end{array} & S_1 & \begin{array}{c}a_3C_1C_2C_3 \\ -a_3C_1S_2S_3 \\ +a_2C_1C_2\end{array} \\[6pt]
\begin{array}{c}S_1C_2C_3 \\ -S_1S_2S_3\end{array} & \begin{array}{c}-S_1C_2S_3 \\ -S_1S_2C_3\end{array} & -C_1 & \begin{array}{c}a_3S_1C_2C_3 \\ -a_3S_1S_2S_3 \\ +a_2S_1C_2\end{array} \\[6pt]
\begin{array}{c}S_2C_3 \\ +C_2S_3\end{array} & C_2C_3 - S_2S_3 & 0 & \begin{array}{c}a_3S_2C_3 \\ +a_3C_2S_3 \\ +a_2S_2 + d_1\end{array} \\[6pt]
0 & 0 & 0 & 1
\end{bmatrix}
$$

$$(4\text{-}80)$$

The position of point P (the arm end in Fig. 4-7) as expressed in frame O_0 is given by the translation vector [see Eq. (4-71)]

$$
\mathbf{d}_0^3 =
\begin{bmatrix} X_p \\ Y_p \\ Z_p \end{bmatrix}
=
\begin{bmatrix}
a_3C_1C_2C_3 - a_3C_1S_2S_3 + a_2C_1C_2 \\
a_3S_1C_2C_3 - a_3S_1S_2S_3 + a_2S_1C_2 \\
a_3S_2C_3 + a_3C_2S_3 + a_2S_2 + d_1
\end{bmatrix}
\quad (4\text{-}81)
$$

Comparing Eqs. (4-81) and (4-5), it is seen that both are identical, excluding d_1 which results from the difference in positioning O_0 relative to O_1 in Figs. 4-2 and 4-7.

Spherical Arm. Consider the spherical arm containing one prismatic and two revolute joints shown in Fig. 4-2b. Following the procedure defined in Sec. 4.4, the links and joints are numerated and links' coordinate systems are assigned. Coordinate system O_3, at the arm end, corresponds to a wrist having its first joint rotating around the Z_3 axis.

The joint parameters established for this case are summarized in

TABLE 4-2. Joint Parameters for the Spherical Arm

Joint i	θ_i	d_i	a_i	α_i
		Joint parameters		
1	θ_1†	0	0	$+90°$
2	θ_2† $+ 90°$	0	0	$+90°$
3	0	d_3†	0	0

†Variables.

Table 4-2. The parameters of joint 1 are similar to those of the articulated arm, but with $d_1 = 0$ in this case. The variable rotation angle of joint 2 is $\theta_2 + 90°$, since it must be measured between X_1 and X_2, while θ_2, as shown in Fig. 4-2b, was chosen differently. Such a modification is required whenever the chosen joint angle cannot be directly used as a joint variable. The variable of joint 3, d_3, describes the translation of frame O_3 along the Z_2 axis.

The three displacement transformation matrices are derived by substituting the joint parameters from Table 4-2 to Eq. (4-72).

$$
\mathbf{T}_0^1 =
\begin{bmatrix}
C_1 & 0 & S_1 & 0 \\
S_1 & 0 & -C_1 & 0 \\
0 & 1 & 0 & 0 \\
0 & 0 & 0 & 1
\end{bmatrix}
$$

$$
\mathbf{T}_1^2 =
\begin{bmatrix}
-S_2 & 0 & C_2 & 0 \\
C_2 & 0 & S_2 & 0 \\
0 & 1 & 0 & 0 \\
0 & 0 & 0 & 1
\end{bmatrix}
\tag{4-82}
$$

$$
\mathbf{T}_2^3 =
\begin{bmatrix}
1 & 0 & 0 & 0 \\
0 & 1 & 0 & 0 \\
0 & 0 & 1 & d_3 \\
0 & 0 & 0 & 1
\end{bmatrix}
$$

Notice that $\mathbf{T}_0^1$ and $\mathbf{T}_1^2$ are actually rotation matrices and $\mathbf{T}_2^3$ is a translation matrix.

The displacement matrix between O_3 and frame O_0 is derived utilizing the chaining operation to the matrices in Eqs. (4-82).

$$
\mathbf{T}_0^3 = \mathbf{T}_0^1 \mathbf{T}_1^2 \mathbf{T}_2^3 =
\left[
\begin{array}{ccc|c}
-C_1S_2 & S_1 & C_1C_2 & d_3C_1C_2 \\
-S_1S_2 & -C_1 & S_1C_2 & d_3S_1C_2 \\
C_2 & 0 & S_2 & d_3S_2 \\
\hline
0 & 0 & 0 & 1
\end{array}
\right]
\tag{4-83}
$$

The position of the TCP in Fig. 4-2b is given by the translation vector

$$
\mathbf{d}_0^3 =
\begin{bmatrix}
X_t \\
Y_t \\
Z_t
\end{bmatrix}
=
\begin{bmatrix}
d_3C_1C_2 \\
d_3S_1C_2 \\
d_3S_2
\end{bmatrix}
\tag{4-84}
$$

which is identical to Eq. (4-3) derived using the straightforward geometry.

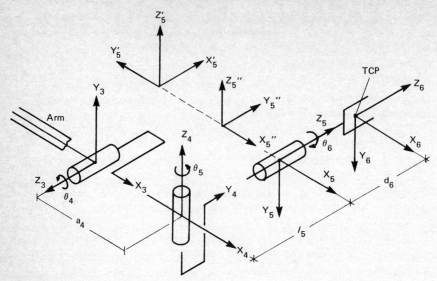

FIG. 4-8 Links' coordinate systems and joint parameters for a bend-bend-roll wrist.

4.5.2 Three-Axis Wrists

Two types of robot wrists will be examined, the bend-bend-roll (BBR) and the roll-bend-roll (RBR) wrist types.

Bend-Bend-Roll Wrist. The BBR wrist is schematically shown in Fig. 4-8 with all joint variables in their reference position. The BBR wrist is used in the T^3 robot (shown in Fig. 1-1) and has pitch-yaw-roll motions. The joints are numerated starting with joint 4 that moves the wrist relative to frame O_3, which is assigned to the arm end. Frame O_6 is assigned to joint 6 (i.e., the end effector) with its origin at the TCP and the Z_6 axis pointing toward the tool direction. The joint parameters for the BBR wrist are summarized in Table 4-3.

TABLE 4-3 Joint Parameters for the BBR Wrist

Joint	Joint parameters			
i	θ_i†	d_i	a_i	α_i
4	θ_4	0	a_4	$-90°$
5	θ_5	‡	‡	$-90°$
6	θ_6	d_6	0	0

†Variables.
‡Undefined.

According to the definitions given in Sec. 4.4, it is impossible to relate the distance l_5 either to a_5 or to d_5 (see Fig. 4-8). The joint parameter a_5, describing the distance between Z_4 and Z_5, should be zero since Z_4 and Z_5 intersect. Moreover, the joint parameter d_5, describing the translation of frame O_5 along axis Z_4, is also zero. Trying to overcome the conflict by choosing an alternative coordinate system for O_5 or O_4 (subject to the obligatory direction of Z_4 and Z_5) is useless in this case.

For such cases, which occur in several joint configurations, the following procedure is suggested: Another coordinate system, O_5', with its axes coinciding (but not matching) with the axes of frame O_5, is assigned to link 5. This frame is chosen so as to enable joint parameters to be defined, from which the displacement matrix between O_4 and O_5' is derived. Then rotation transformation matrices are used to transform O_5' to O_5. Subsequently, the transformation matrix between O_4 and O_5 is established using the chaining operation. For this example (Fig. 4-8), a coordinate system O_5' was chosen and the parameters of joint 5 were defined as $\theta_5' = \theta_5 + 90°$, $d_5' = 0$, $a_5' = l_5$, $\alpha_5' = 0$, yielding a displacement transformation matrix.

$$\mathbf{T}_4^{5'} = \begin{bmatrix} -\sin\theta_5 & -\cos\theta_5 & 0 & -l_5\sin\theta_5 \\ \cos\theta_5 & -\sin\theta_5 & 0 & l_5\cos\theta_5 \\ 0 & 0 & 1 & 0 \\ 0 & 0 & 0 & 1 \end{bmatrix} \tag{4-85}$$

Now, two consecutive rotations of O_5' define the transformation matrix between O_5 and O_5'. First O_5' is rotated around Z_5' axis, using Eq. (4-61) with $\theta = -90°$, to form O_5'' frame. Then O_5'' is rotated around X_5'' axis, using Eq. (4-59) with $\alpha = -90°$ to form the O_5 frame

$$\mathbf{T}_{5'}^5 = \mathbf{T}_{R5'}^{5''}\mathbf{T}_{R5''}^5 = \begin{bmatrix} 0 & 1 & 0 & 0 \\ -1 & 0 & 0 & 0 \\ 0 & 0 & 1 & 0 \\ 0 & 0 & 0 & 1 \end{bmatrix} \begin{bmatrix} 1 & 0 & 0 & 0 \\ 0 & 0 & 1 & 0 \\ 0 & -1 & 0 & 0 \\ 0 & 0 & 0 & 1 \end{bmatrix}$$

$$= \begin{bmatrix} 0 & 0 & 1 & 0 \\ -1 & 0 & 0 & 0 \\ 0 & -1 & 0 & 0 \\ 0 & 0 & 0 & 1 \end{bmatrix} \tag{4-86}$$

The transformation between O_4 and O_5 frame is

$$\mathbf{T}_4^5 = \mathbf{T}_4^{5'}\mathbf{T}_{5'}^5 \tag{4-87}$$

The validity of Eq. (4-86) can be checked by

$$
\begin{bmatrix} X_5' \\ Y_5' \\ Z_5' \\ 1 \end{bmatrix} = T_{5'}^5 \begin{bmatrix} X_5 \\ Y_5 \\ Z_5 \\ 1 \end{bmatrix} = \begin{bmatrix} Z_5 \\ -X_5 \\ -Y_5 \\ 1 \end{bmatrix}
\tag{4-88}
$$

Namely, Z_5 is in the direction of X_5'; $-X_5 \equiv Y_5'$; and $-Y_5 \equiv Z_5'$.
The transformation matrices for the BBR wrist are

$$
T_3^4 = \begin{bmatrix}
\cos \theta_4 & 0 & -\sin \theta_4 & a_4 \cos \theta_4 \\
\sin \theta_4 & 0 & \cos \theta_4 & a_4 \sin \theta_4 \\
0 & -1 & 0 & 0 \\
0 & 0 & 0 & 1
\end{bmatrix}
$$

$$
T_4^5 = \begin{bmatrix}
\cos \theta_5 & 0 & -\sin \theta_5 & -l_5 \sin \theta_5 \\
\sin \theta_5 & 0 & \cos \theta_5 & l_5 \cos \theta_5 \\
0 & -1 & 0 & 0 \\
0 & 0 & 0 & 1
\end{bmatrix}
\tag{4-89}
$$

$$
T_5^6 = \begin{bmatrix}
\cos \theta_6 & -\sin \theta_6 & 0 & 0 \\
\sin \theta_6 & \cos \theta_6 & 0 & 0 \\
0 & 0 & 1 & d_6 \\
0 & 0 & 0 & 1
\end{bmatrix}
$$

Observing Eqs. (4-89) it is seen that it is impossible to find joint variables for deriving T_4^5 out of Eq. (4-72) according to the DH convention.
The displacement transformation matrix between the TCP and the edge of the robot arm for the BBR wrist is

$$
T_3^6 = T_3^4 T_4^5 T_5^6
\tag{4-90}
$$

Roll-Bend-Roll Wrist.   The RBR wrist, which is commonly used in robot manipulators (e.g., the PUMA with $d_4 = l_5 = 0$), is illustrated in Fig. 4-9. All joint variables are shown in their reference position.
The joint parameters for this wrist are summarized in Table 4-4. The transformation matrix between O_4 and O_5 is identical to the associated matrix established for the BBR wrist by means of two successive transformations.
The transformation matrices for the RBR wrist are

$$
T_3^4 = \begin{bmatrix}
\cos \theta_4 & 0 & \sin \theta_4 & 0 \\
\sin \theta_4 & 0 & -\cos \theta_4 & 0 \\
0 & 1 & 0 & d_4 \\
0 & 0 & 0 & 1
\end{bmatrix}
$$

$$\mathbf{T}_4^5 = \begin{bmatrix} \cos\theta_5 & 0 & -\sin\theta_5 & l_5\sin\theta_5 \\ \sin\theta_5 & 0 & \cos\theta_5 & l_5\cos\theta_5 \\ 0 & -1 & 0 & 0 \\ 0 & 0 & 0 & 1 \end{bmatrix} \qquad (4\text{-}91)$$

$$\mathbf{T}_5^6 = \begin{bmatrix} \cos\theta_6 & -\sin\theta_6 & 0 & 0 \\ \sin\theta_6 & \cos\theta_6 & 0 & 0 \\ 0 & 0 & 1 & d_6 \\ 0 & 0 & 0 & 1 \end{bmatrix}$$

TABLE 4-4 Joint Parameters for the RBR Wrist

Joint i	Joint parameters			
	θ_i†	d_i	a_i	α_i
4	θ_4	d_4	0	$+90°$
5	θ_5	‡	‡	$-90°$
6	θ_6	d_6	0	0

†Joint variables.
‡Undefined.

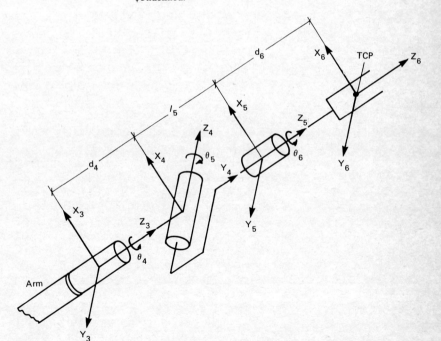

FIG. 4-9 Links' coordinate systems and joint parameters for a roll-bend-roll wrist.

The displacement transformation matrix between the TCP and the arm end is given by Eq. (4-90) for the matrices of Eqs. (4-91).

4.5.3 Six-Axis Robot Manipulators

Since various types of arms and wrists can be combined to form a manipulator, the following will demonstrate the procedure to merge the transformation matrices of an arm and a wrist. Transformation matrices, derived for the two types of three-axis wrists, will be chained with the transformation matrices derived for the two arm types to form the transformation matrix for a six-axis manipulator.

Articulated Arm with BBR Wrist.   A six-axis manipulator is constructed by combining the BBR wrist in Fig. 4-8 to the articulated arm in Fig. 4-7. The transformation matrix for the resultant six-axis manipulator is derived by chaining the transformation matrices of the wrist with those of the arm. Since frame O_3 in Fig. 4-8 matches frame O_3 in Fig. 4-7, they can be united, and accordingly, the chaining operation is enabled:

$$\mathbf{T}_0^6 = \mathbf{T}_0^3 \mathbf{T}_3^6 \qquad (4\text{-}92)$$

where $\mathbf{T}_0^3$ is given in Eq. (4-80) for the articulated arm, $\mathbf{T}_3^6$ is given in Eq. (4-90) for the BBR wrist, and $\mathbf{T}_0^6$ is the transformation matrix for the resultant manipulator. The position of the TCP in the WCS (O_0), $\mathbf{d}_0^6$, and the orientation of the end effector, $\mathbf{C}_0^6$, are derived from $\mathbf{T}_0^6$ according to Eq. (4-71).

Spherical Arm with BBR Wrist.   The BBR wrist in Fig. 4-8 can also be combined with a spherical arm in Fig. 4-2b to form a six-axis manipulator. The chaining of the transformation matrices of the wrist and the arm cannot be performed directly from Eq. (4-92) since, in contrast to the previous case, the two frames O_3 in Figs. 4-8 and 4-2b do not match. In order to enable the chaining operation, an addition transformation matrix between frame O_3 in Fig. 4-2b and O_3 in Fig. 4-8 is required.

$$\mathbf{T}_a^w = \begin{bmatrix} 0 & 1 & 0 & 0 \\ 0 & 0 & 1 & 0 \\ 1 & 0 & 0 & 0 \\ 0 & 0 & 0 & 1 \end{bmatrix} \qquad (4\text{-}93)$$

where w and a denote the wrist and the arm, respectively. The matrix $\mathbf{T}_a^w$ (obtained by rotating O_3 in Fig. 4-8 with respect to O_3 in Fig. 4-2b) is used to transform point coordinates from frame O_3 of the wrist to frame O_3 of the arm according to

$$\begin{bmatrix} X \\ Y \\ Z \\ 1 \end{bmatrix}_a = \mathbf{T}_a^w \begin{bmatrix} X \\ Y \\ Z \\ 1 \end{bmatrix}_w = \begin{bmatrix} Y \\ Z \\ X \\ 1 \end{bmatrix}_w \tag{4-94}$$

Now, the chaining operation is performed according to

$$\mathbf{T}_0^6 = \mathbf{T}_0^3 \mathbf{T}_a^w \mathbf{T}_3^6 \tag{4-95}$$

where $\mathbf{T}_0^3$ is given in Eq. (4-83), $\mathbf{T}_a^w$ is given in Eq. (4-93), $\mathbf{T}_3^6$ is given in Eq. (4-90), and $\mathbf{T}_0^6$ is the transformation matrix for the resultant manipulator.

Note that the effect of $\mathbf{T}_a^w$ is in replacing the O_3 frame in Fig. 4-2b by another coordinate system, and consequently replacing $\mathbf{T}_2^3$ in Eqs. (4-82) by

$$\mathbf{T}_2^3 \mathbf{T}_a^w = \begin{bmatrix} 0 & 1 & 0 & 0 \\ 0 & 0 & 1 & 0 \\ 1 & 0 & 0 & d_3 \\ 0 & 0 & 0 & 1 \end{bmatrix} \tag{4-96}$$

with which Eq. (4-92) can be used instead of Eq. (4-95).

Arms with RBR Wrists.   Combining the RBR wrist with the spherical arm is performed according to Eq. (4-92) since the two frames O_3 in both the arm and the wrist coincide.

Combining the RBR wrist with the articulated arm shown in Fig. 4-7 (forming a PUMA-type manipulator) is performed according to Eq. (4-95) with

$$\mathbf{T}_a^w = \begin{bmatrix} 0 & 0 & 1 & 0 \\ 1 & 0 & 0 & 0 \\ 0 & 1 & 0 & 0 \\ 0 & 0 & 0 & 1 \end{bmatrix} \tag{4-97}$$

and for $\mathbf{T}_0^3$ and $\mathbf{T}_3^6$ given in Eqs. (4-80) and (4-91), respectively.

4.5.4 Assigning the Tool Coordinate System

Recalling the examples of the two wrist types, the frame O_6 was assigned to the end effector with its origin at the TCP, and its Z_6 axis pointing to the direction of the end effector. Such systems are frequently denoted as TCS.

In the two TCS shown in Figs. 4-8 and 4-9, the Z_6 axis and the TCP were illustrated coinciding with the Z_5 axis. However, there are cases in which the direction of the Z_6 axis (dictated by the orientation of the end effector relative to the edge of frame O_5) and the location of the TCP demonstrate a different situation. For these cases, a transformation matrix between O_6 and O_5 should be derived, based upon the parameters used in the programming language to describe the end effector. These parameters (that are dependent on the programming language) are used by the operator to describe the position and orientation of the end effector with respect to the wrist end.

For example, consider a programming language that uses three parameters to describe the location of the TCP relative to the center of a flange mounted at the end of the wrist. Assuming an RBR wrist (illustrated in Fig. 4-9), these three parameters are defined in the programming language relative to a reference coordinate system which is assigned to the center of the flange and its Z axis coinciding with the Z_5 axis. In this case the O_5 frame can serve as the reference coordinate system, with $\theta_6 = 0$ and with the joint parameter l_5 chosen to bring the origin O_5 to the flange edge. The three parameters (ε_1, ε_2, and ε_3) describe the coordinates of the TCP in X_5, Y_5, and Z_5, respectively.

It is now necessary to derive the transformation matrix between O_6', located at the TCP, and frame O_5. This matrix should be a function of the three parameters (ε_1, ε_2, and ε_3) and of the angle θ_6. Referring back to Fig. 4-9, this matrix can be obtained by two successive transformations

$$\mathbf{T}_5^{6'} = \mathbf{T}_5^6 \begin{bmatrix} 1 & 0 & 0 & \varepsilon_1 \\ 0 & 1 & 0 & \varepsilon_2 \\ 0 & 0 & 1 & \varepsilon_3 \\ 0 & 0 & 0 & 1 \end{bmatrix} \tag{4-98}$$

where $\mathbf{T}_5^6$ is given in Eqs. (4-91) with $d_6 = 0$, and the second matrix translates O_6' relative to O_6 frame. It can be shown that after matrix multiplication and arranging of terms, Eq. (4-98) is of the form

$$\mathbf{T}_5^{6'} = \begin{bmatrix} \cos \theta_6 & -\sin \theta_6 & 0 & a_6 \cos \phi_6 \\ \sin \theta_6 & \cos \theta_6 & 0 & a_6 \sin \phi_6 \\ 0 & 0 & 1 & d_6 \\ 0 & 0 & 0 & 1 \end{bmatrix} \tag{4-99}$$

where $a_6 = (\varepsilon_1^2 + \varepsilon_2^2)^{1/2}$
$\phi_6 = \cos^{-1}(\varepsilon_1/a_6) + \theta_6$
$d_6 = \epsilon_3$

For programming languages with which it is also possible to define the orientation of the end effector relative to the flange, the transformation matrix between the TCS and frame O_5 is

$$
\mathbf{T}_5^{6'} =
\begin{bmatrix}
\cos\theta_6 & -\sin\theta_6 & 0 & a_6\cos\phi_6 \\
\sin\theta_6 & \cos\theta_6 & 0 & a_6\sin\phi_6 \\
0 & 0 & 1 & d_6 \\
0 & 0 & 0 & 1
\end{bmatrix}
\begin{bmatrix}
 & & & 0 \\
 & \mathbf{C}^* & & 0 \\
 & & & 0 \\
0 & 0 & 0 & 1
\end{bmatrix}
\tag{4-100}
$$

where $\mathbf{C}^*$ describes the orientation of the end effector (or tool) relative to the flange of the wrist. Since the above-mentioned orientation matrix is the DCM, the programming language should require the definition of its three independent parameters. Alternatively, the programming language can utilize other parameters, which are more comprehensive to the operator, with which $\mathbf{C}^*$ can be constructed indirectly. Such representation can be based upon the Euler angles or the rotation vector presented in Sec. 4.6.

Note that in practice it is convenient to separate between $\mathbf{T}_5^6$ and $\mathbf{T}_5^{6'}$, since the former is constant for a given robot and the latter is dependent on the end effector.

4.6 QUATERNION AND ROTATION VECTOR REPRESENTATIONS

In the previous sections the DCM was utilized to define the relative orientation of two coordinate systems. An alternative method to define this relation and to perform coordinate transformation is by using a quaternion or a rotation vector. The advantages of these methods are that the definition of the relative orientation consists of three or four parameters instead of nine (3×3 matrix) as in the DCM definition. Moreover, the number of the arithmetic operations required for the kinematic solution of a robot manipulator can be decreased using the quaternion or rotation vector methods.

The relative orientation representations by the quaternion and the rotation vector are based on the Euler theorem which states that a displacement of a rigid body with one fixed point can be described as a rotation about some axis (Begges, 1966; Duffy, 1980). The rotation vector is a vector pointing along this axis, and its magnitude contains information about the rotation angle. Consequently, the rotation vector is the only common vector of any two coordinate systems (i.e., frames) having the same origin and differing in their orientation. This feature is applied to perform vector transformation between the two frames. The direction of the rotation vector defines the axis around which one coordinate system has to be rotated to achieve the same orientation as the other system, while its magnitude defines the amount of rotation. Utilizing the rotation vector one can define the relative orientation of two systems and perform vector transformation from one coordinate system to another.

The quaternion is a four-element vector, three of which are components of a spatial vector and define the direction of the common rotation axis (i.e., coincides with the direction of the rotation vector), and the fourth component provides information about the rotation angle.

Two implementations of the quaternion and the rotation vector will be discussed; vector transformation between two frames and a representation of successive rotations of frames by a single rotation.

4.6.1 Quaternion Definition

The quaternion q has four elements which completely identify the direction of the axis of rotation and the amount of rotation. It consists of scalar element q_0 and a vector $\mathbf{q}$ and can be written as:

$$q = q_0 + \mathbf{q} = q_0 + q_1\hat{\mathbf{i}} + q_2\hat{\mathbf{j}} + q_3\hat{\mathbf{k}} \tag{4-101}$$

or $\qquad q = (q_0, \mathbf{q})$

where $\hat{\mathbf{i}}$, $\hat{\mathbf{j}}$, and $\hat{\mathbf{k}}$ are unit vectors along the X, Y, and Z axes, respectively. It is frequently convenient to represent a vector as a quaternion with a scalar element equal to zero.

The quaternion which is used to identify relative rotation by an angle θ about an axis, whose direction is defined by the unit vector $\hat{\mathbf{e}}$, was defined by Euler as

$$q = \cos\frac{\theta}{2} + \hat{\mathbf{e}}\sin\frac{\theta}{2} \tag{4-102}$$

or

$$q = \cos\frac{\theta}{2} + (e_1\hat{\mathbf{i}} + e_2\hat{\mathbf{j}} + e_3\hat{\mathbf{k}})\sin\frac{\theta}{2}$$

where e_1, e_2, and e_3 are the direction cosines of $\hat{\mathbf{e}}$. Alternatively the quaternion can be written as

$$q = q_0 + \mathbf{q} = \cos\frac{\theta}{2} + \begin{bmatrix} e_1 \\ e_2 \\ e_3 \end{bmatrix}\sin\frac{\theta}{2} \tag{4-103}$$

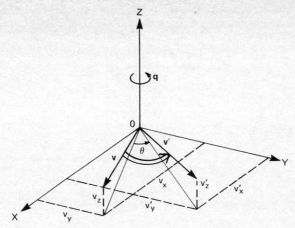

FIG. 4-10   Rotation of a vector about the Z axis.

4.6.2 Vector Transformation Using the Quaternion

Assume that a vector **v** in frame O undergoes a rotation described by the quaternion q in frame O. The resulting vector in frame O is **v′**, and its relation to **v** is given by the equation (Hillman, 1964; Shoham, 1982; Wiener, 1962; Whipple, 1971)

$$\mathbf{v}' = \mathbf{v} + 2q_0(\mathbf{q} \times \mathbf{v}) + 2\mathbf{q} \times (\mathbf{q} \times \mathbf{v}) \tag{4-104}$$

Equation (4-104) is also interpreted as a coordinate transformation, which means a description of the vector components in two frames with different orientations. (The vector is fixed in space and does not rotate as previously.) In Eq. (4-104), the fixed vector **v** in frame O is seen as **v′** in frame O'; the relation between the two frames is given by the quaternion q in O'.

As an example, consider a vector **v** in frame O as shown in Fig. 4-10, which undergoes a rotation defined by q. The axis of rotation coincides with the Z axis, thus the quaternion definition of this rotation according to Eq. (4-103) is

$$q = \cos\frac{\theta}{2} + \begin{bmatrix} 0 \\ 0 \\ 1 \end{bmatrix} \sin\frac{\theta}{2}$$

where θ is the angle of rotation around the Z axis.

The vector rotation results in a vector **v′** in frame O and is given according to Eq. (4-104)

$$
\begin{bmatrix} v'_x \\ v'_y \\ v'_z \end{bmatrix} = \begin{bmatrix} v_x \\ v_y \\ v_z \end{bmatrix} + 2\cos\frac{\theta}{2} \begin{bmatrix} 0 \\ 0 \\ 1 \end{bmatrix} \times \begin{bmatrix} v_x \\ v_y \\ v_z \end{bmatrix} \sin\frac{\theta}{2}
$$

$$
+ 2 \begin{bmatrix} 0 \\ 0 \\ 1 \end{bmatrix} \times \left(\begin{bmatrix} 0 \\ 0 \\ 1 \end{bmatrix} \times \begin{bmatrix} v_x \\ v_y \\ v_z \end{bmatrix} \right) \sin^2\frac{\theta}{2}
$$

$$
= \begin{bmatrix} v_x \\ v_y \\ v_z \end{bmatrix} + 2\cos\frac{\theta}{2}\sin\frac{\theta}{2} \begin{bmatrix} -v_y \\ v_x \\ 0 \end{bmatrix} + 2\sin^2\frac{\theta}{2} \begin{bmatrix} -v_x \\ -v_y \\ 0 \end{bmatrix}
$$

$$
= \begin{bmatrix} v_x \cos\theta - v_y \sin\theta \\ v_y \cos\theta + v_x \sin\theta \\ v_z \end{bmatrix}
$$

This result is identical with the one given in Eq. (4-10).

4.6.3 Successive Rotations

Two successive rotations which are described by two quaternions g and g' can be represented as a single rotation described by one quaternion according to the following equation (Ickes, 1970; Shoham, 1982; Whittaker, 1944)

$$
q = gg' = (g_0, \mathbf{g})(g'_0, \mathbf{g}') \tag{4-105}
$$
$$
= (g_0 g'_0 - \mathbf{g}\cdot\mathbf{g}', \, g_0\mathbf{g}' + g'_0\mathbf{g} + \mathbf{g} \times \mathbf{g}')
$$

Each of the quaternions g and g' in the above equation is resolved in its local coordinate system in which the rotation is taking place.

As an example, consider that the vector $\mathbf{v}$ in Fig. 4-11 undergoes two successive rotations. The first rotation is defined by the quaternion g, rotates the vector $\mathbf{v}$ and a reference coordinate system O' attached to $\mathbf{v}$ (originally O' coincides with O) through an angle of 90° about the Z axis, and creates the vector $\mathbf{v}'$. The second rotation, defined by the quaternion g', in the O' frame, rotates the vector $\mathbf{v}'$ and a frame O'' attached to it through 90° about the X' axis into the final position. Let us find the components of $\mathbf{v}$ at its final position ($\mathbf{v}''$), if it is described originally as $\mathbf{v} = [1, 0, 0]^T$.

The solution is divided into two parts. The first describes the two successive rotations as a single composite rotation, and the second rotates the vector $\mathbf{v}$ through this rotation to find the final position. To find the composite rotation q, each of the quaternions g and g' has to be defined in its local coordinate system.

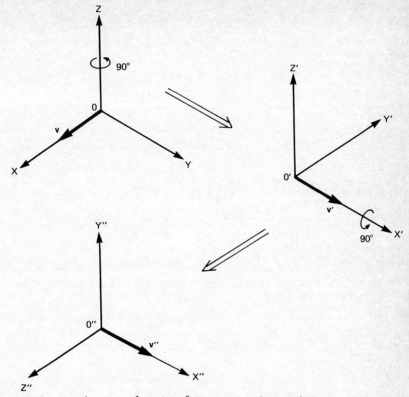

FIG. 4-11 The vector direction after two successive rotations.

The quaternion g relates the O and the O' frames and is defined according to Eq. (4-102). For a rotation of $\theta = 90°$ around the Z axis

$$g = (g_0, \mathbf{g}) = \left(\frac{1}{\sqrt{2}}, 0\hat{\mathbf{i}} + 0\hat{\mathbf{j}} + \frac{1}{\sqrt{2}}\hat{\mathbf{k}} \right)$$

Similarly, g' relates the O' and the O'' frames and is defined by a 90° rotation around the X' axis

$$g' = \left(g_0', \mathbf{g}' \right) = \left(\frac{1}{\sqrt{2}}, \frac{1}{\sqrt{2}}\hat{\mathbf{i}}' + 0\hat{\mathbf{j}}' + 0\hat{\mathbf{k}}' \right)$$

The composite rotation, q, is obtained according to Eq. (4-105)

$$q = gg' = (½, ½\hat{\mathbf{i}} + ½\hat{\mathbf{j}} + ½\hat{\mathbf{k}}) \tag{4-106}$$

Note that q, the composite rotation, is given in frame O, and relates the O'' to O frame.

The next step in the solution is to rotate the vector **v** through the composite rotation obtained in Eq. (4-106) by using Eq. (4-104).

$$\mathbf{v}'' = \begin{bmatrix} 1 \\ 0 \\ 0 \end{bmatrix} + 2(\tfrac{1}{2}) \begin{bmatrix} \tfrac{1}{2} \\ \tfrac{1}{2} \\ \tfrac{1}{2} \end{bmatrix} \times \begin{bmatrix} 1 \\ 0 \\ 0 \end{bmatrix} + 2 \begin{bmatrix} \tfrac{1}{2} \\ \tfrac{1}{2} \\ \tfrac{1}{2} \end{bmatrix}$$

$$\times \left(\begin{bmatrix} \tfrac{1}{2} \\ \tfrac{1}{2} \\ \tfrac{1}{2} \end{bmatrix} \times \begin{bmatrix} 1 \\ 0 \\ 0 \end{bmatrix} \right) = \begin{bmatrix} 0 \\ 1 \\ 0 \end{bmatrix} \qquad (4\text{-}107)$$

The result of Eq. (4-107) shows that a unit vector along the X axis becomes, after the two rotations, a unit vector along the Y axis, as one can verify in Fig. 4-11. Also note that the transformation from **v** to **v**″ can be obtained by a single rotation of $\theta = 120°$ about an axis which has equal projections on the X, Y, and Z axes.

4.6.4 Rotation Vector

An alternative representation of a rotation is by using a three-dimensional vector instead of the four-element quaternion. The basis of this representation is similar to the one presented for the quaternion and consists of the existence of a common axis around which one coordinate system has to be rotated to achieve the same orientation as another coordinate system. A vector along this common axis, whose magnitude is defined by the amount of rotation around this axis, is called the *rotation vector*. There are several definitions for the magnitude of the rotation vector (Shoham, 1982; Wiener, 1962); however, we prefer to use the following definition for a rotation vector **R**:

$$\mathbf{R} = \hat{\mathbf{e}} \tan \frac{\theta}{2} = \hat{\mathbf{e}} f \qquad (4\text{-}108)$$

where $\hat{\mathbf{e}}$ = unit vector along axis of rotation
θ = angle of rotation $\hat{\mathbf{e}}$
$f = \tan \theta/2$ = magnitude of rotation vector

The rotation vector can be used, similarly to the quaternion, for vector transformation and in applying a single rotation representation for several rotations, as will be explained below.

A vector **v** undergoing a rotation **R** to perform a vector **v**′ can be described as

$$\mathbf{v}' = \mathbf{v} + \frac{2\mathbf{R} \times (\mathbf{v} + \mathbf{R} \times \mathbf{v})}{1 + f^2} \qquad (4\text{-}109)$$

Two successive rotations, $\mathbf{R}_1$ and $\mathbf{R}_2$, which are resolved in their local coordinate systems, can be combined to form one composite rotation by

$$\mathbf{R} = \mathbf{R}_1 \circ \mathbf{R}_2 = \frac{\mathbf{R}_1 + \mathbf{R}_2 + \mathbf{R}_1 \times \mathbf{R}_2}{1 - \mathbf{R}_1 \cdot \mathbf{R}_2} \qquad (4\text{-}110)$$

where the symbol $\circ$ describes a rotation operation acting on rotation vector as defined in the above equation. Equations (4-109) and (4-110) are derived by substituting $\mathbf{q} = \mathbf{R} \cos \theta/2$ into Eqs. (4-104) and (4-105), respectively.

The vector representation of rotation requires only three elements rather than four as in the quaternion. Although there is a redundancy in the four-element representation, it eliminates the singularity problem which arises in the rotation vector method corresponding to a half or a full revolution ($\theta = n\pi, n = 1, 2, \ldots$).

As an example of the rotation vector method, consider a spherical robot holding a tool in its end effector. It is desired to resolve the tool vector coordinates in the base frame. Obviously, the tool orientation is changed because of the motion of the robot's joints.

The robot configuration and the joint variables are shown in Fig. 4-12. The revolute joint variables θ_1 and θ_2 change the tool orientation,

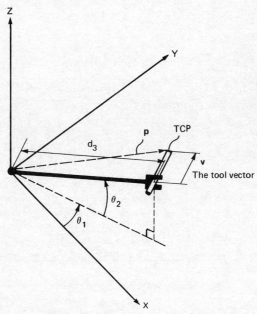

FIG. 4-12 The coordinates of a tool handled by a spherical robot.

and consequently the tool vector coordinates vary as well. However, the prismatic joint variable d_3 does not change the tool vector coordinates since it translates the tool in parallel to its previous location.

The solution consists of describing the two rotations, θ_1 and θ_2, as a single composite rotation, and transforming the tool vector from its initial position to its final position by this composite rotation.

According to Fig. 4-12, the first rotation is around the Z axis in the robot base frame and therefore the first rotation vector is

$$\mathbf{R}_1 = [0, 0, f_1]^T$$

The next rotation is elevation about the shifted Y axis, and the corresponding rotation vector is

$$\mathbf{R}_2 = [0, -f_2, 0]^T$$

where $f_1 = \tan \theta_1/2$ and $f_2 = \tan \theta_2/2$.

The composite rotation $\mathbf{R}$ is given according to Eq. (4-110)

$$\mathbf{R} = \frac{\begin{bmatrix} 0 \\ 0 \\ f_1 \end{bmatrix} + \begin{bmatrix} 0 \\ -f_2 \\ 0 \end{bmatrix} + \begin{bmatrix} 0 \\ 0 \\ f_1 \end{bmatrix} \times \begin{bmatrix} 0 \\ -f_2 \\ 0 \end{bmatrix}}{1 - \begin{bmatrix} 0 \\ 0 \\ f_1 \end{bmatrix} \begin{bmatrix} 0 \\ -f_2 \\ 0 \end{bmatrix}} = \begin{bmatrix} f_1 f_2 \\ -f_2 \\ f_1 \end{bmatrix}$$

For simplicity, assume that the tool in Fig. 4-12 is initially parallel to the Y axis and the tool vector coordinates are initially $\mathbf{v} = [0, 2, 0]^T$. In the initial position the joint variables θ_1 and θ_2 are zero, and at the final position θ_1 and θ_2 are 90° and 60°, respectively.

To reach the final position, the robot's tool performs the rotation $\mathbf{R}$ and the final tool vector $\mathbf{v}'$ becomes, according to Eq. (4-109),

$$\mathbf{v}' = \begin{bmatrix} 0 \\ 2 \\ 0 \end{bmatrix} + \frac{2 \begin{bmatrix} f_1 f_2 \\ -f_2 \\ f_1 \end{bmatrix} \times \left(\begin{bmatrix} 0 \\ 2 \\ 0 \end{bmatrix} + \begin{bmatrix} f_1 f_2 \\ -f_2 \\ f_1 \end{bmatrix} \times \begin{bmatrix} 0 \\ 2 \\ 0 \end{bmatrix} \right)}{1 + f_1^2 f_2^2 + f_2^2 + f_1^2}$$

$$= \begin{bmatrix} 0 \\ 2 \\ 0 \end{bmatrix} + \frac{-4f_1}{(f_1^2 + 1)} \begin{bmatrix} 1 \\ f_1 \\ 0 \end{bmatrix} \tag{4-111}$$

For the given angle of rotations

$$f_1 = \tan \frac{\theta_1}{2} = \tan \frac{90°}{2} = 1$$

$$f_2 = \tan \frac{\theta_2}{2} = \tan \frac{60°}{2} = \frac{1}{\sqrt{3}}$$

Note that neither f_2 nor d_3 appear in Eq. (4-111) because the tool axis is parallel to the θ_2 axis of rotation and because of the parallel motion of d_3.

Solving the tool vector coordinates at the tool's final position,

$$\mathbf{v} = \begin{bmatrix} 0 \\ 2 \\ 0 \end{bmatrix} + \frac{-4.1}{1 + 1} \begin{bmatrix} 1 \\ 1 \\ 0 \end{bmatrix} = \begin{bmatrix} -2 \\ 0 \\ 0 \end{bmatrix}$$

The results indicate that the tool vector, which coincides originally with the Y axis, points now to the negative X direction.

It must be noted that in this procedure of vector transformation only the tool vector coordinates in the base frame are resolved, but not the TCP position (see Fig. 4-12). This can be easily deduced from Eq. (4-111) in which $\mathbf{v}'$ is a function of θ_1 alone but not of θ_2 and d_3. The robot TCP, however, depends on all three joint variables. In order to derive the TCP in the base frame, one has to perform the same rotation $\mathbf{R}$ acting on a vector which is drawn from the origin of the base frame toward the TCP (vector $\mathbf{p}$ in Fig. 4-12), instead of performing the rotation on the tool vector itself. This is left as an exercise to the reader.

REFERENCES

Begges, J.S.: *Advanced Mechanism*, Macmillan Co., New York, 1966.

Brady, M., J. M. Hollerbach, T. L. Johnson, T. Lazano-Perez, and M. T. Mason: *Robot Motion: Planning and Control*, MIT Press, Cambridge, Mass., 1982.

Denavit, J., and R. S. Hartenberg: "A Kinematic Notation for Lower Pair Mechanisms Based on Matrices," *J. Appl. Mech. ASME*, June 1955, pp. 215–221.

Duffy, J.: *Analysis of Mechanisms and Robot Manipulators*, John Wiley & Sons, New York, 1980.

Elord, B. D.: "Some Properties of Quaternions Related to Euler's Theorem," Bellcomm Inc., B70 06076, 1970.

Hillman, A: "Euler Symmetric Parameters," *Isr. Annu. Conf. Aviat. Astronaut.*, January 1964, pp. 125–130.

Hollingsworth, C.A.: *Vector Matrices and Group Theory for Scientists and Engineers*, McGraw-Hill, New York, 1967.

Ickes, B. P.: "A New Method for Performing Digital Control System Attitude Computation Using Quaternions," *AIAA J.*, vol. 8, no. 1, January 1970, pp. 13–17.

Lee, G.: *Robotics Theory and Practice*, to be published.

Lien, K.: "Coordinate Transformation in CNC Systems for Automatic Handling Machine," *Proc. CIRP Semin. Manuf. Syst.*, vol. 9, no. 1, 1980, pp. 49–60.

Mayo, R. A.: "Relative Quaternion State Transition Rate," *J. Guid. Control*, vol. 2, no. 1, 1977, pp. 44–48.

Morecki, A., and K. Kedzior: *Theory and Practice of Robots and Manipulators*, PWN, Polish Scientific Publishers, 1976.

Paul, R.: *Robot Manipulators: Mathematics, Programming, and Control*, MIT Press, Cambridge, Mass., 1981.

Pieper, D. L.: "The Kinematics of Manipulators under Computer Control," Ph.D. Thesis, Stanford University, 1969.

Renaud, M.: "Contribution à la Modelisation et à la Commande Dynamique des Robots Manipulateurs," Ph. D. Thesis, Université Paul Sabatier de Toulouse, September 1980.

Shepperd, S. W.: "Quaternion from Rotation Matrix," *J. Guid. Control*, vol. 1, no. 3, 1978, pp. 223–224.

Shoham, M.: "Development of Algorithms for Robot Spatial Motions," M.Sc. Thesis, Technion, Israel Institute of Technology, April 1982 (in Hebrew).

Suh, C. H., and C. W. Radcliffe: *Kinematics and Mechanisms Design*, John Wiley & Sons, New York, 1978.

Whipple, P. H.: "The Quarternion Representation of Rotations and Its Application in the Skylab Orbital Assembly," Bellcomm Inc., B71 03078, March 1971.

Whittaker, E. T.: *A Treatise on the Analytical Dynamics of Particles and Rigid Bodies*, Dover Publications, New York, 1944.

Wiener, T. F.: "Theoretical Analysis of Gimballess Inertial Reference Equipment Using Delta-Modulated Instruments," Ph.D. Thesis, MIT, Cambridge, Mass., March 1962.

CHAPTER 5

Trajectory Interpolators†

The solution of the direct kinematics problem, which determines the state of the end effector from the known states of the various axes of motion, was described in Chap. 4. The present chapter deals with the solution of the inverse kinematics problem, namely determining the state of the axes of motion from the given state of the end effector in space. The inverse kinematics is implemented in the *trajectory interpolator* algorithm of CP robots. In these robots the interpolator coordinates the motion commands to the individual axes according to the instantaneous desired state of the end effector.

This chapter examines several inverse kinematics techniques and their resulting interpolator algorithm. Section 5.1 introduces the principle of trajectory interpolation in CP robots and the associated problems arising in solving the inverse kinematics problem. It is followed by the derivation of the inverse kinematics solution for particular cases (Sec. 5.2). The rest of the chapter describes two systematic techniques to the solution of the problem. Section 5.3 presents the *resolved motion rate control method*, which is a general technique that can be implemented in any joint configuration and is almost independent of the complexity of the mechanism under consideration. This technique suits an open kinematic chain manipulator with six axes of motion, but becomes impractical when the manipulator has four or five axes. The *quaternion technique*, presented in Sec. 5.4, mainly suits a manipulator

---

†This chapter was contributed by G. Amitai.

consisting of six axes of motion with the three axes of the wrist intersecting at one point. This technique requires a fewer number of arithmetic operations and shows better trajectory-tracing characteristics. There are other techniques such as the approximated step method (Amitai, 1983) which suits mainly four- or five-axis manipulators that can also be driven by closed kinematic chains (e.g., four-bar mechanisms). Such techniques, however, are not as well known as the first two and are not presented in this text.

5.1 INTRODUCTION

Motion commands in CP robots can be defined along the axes of several useful coordinate systems, as shown in Fig. 5-1. The WCS is referred to the stationary base of the manipulator (and thus it describes absolute coordinates), the TCS is referred to the end effector (or to a sensor mounted on the wrist), and the SCS is assigned to a sensor mounted above the working volume of the robot. Although Fig. 5-1 uses cartesian coordinates, cylindrical or spherical coordinate systems can also be used.

In robotics, motions defined in any of the above-mentioned coordinate systems are related either to pure rotation about or pure displacement along one of the coordinate axes. With a displacement command the orientation of the end effector remains unchanged in the associated coordinate system, and with a rotation command the absolute position of the TCP is not changed.

5.1.1 The Necessity of Interpolators

Trajectory interpolators are required in both the teaching and the operating modes of CP robots. Applying joint commands in manual teaching and offline programming of these robots is impractical, and therefore either the WCS or the TCS is used (see Chap. 7).

The more useful coordinate reference in manual teaching mode is the TCS. In this mode, the displacement commands—reach, lift, and sweep—and the rotation commands—pitch, yaw, and roll (see Fig. 5-1)—refer to the current position of the TCP and the direction toward which the end effector is pointing; thus they are easily understood by the operator. When the TCS and WCS are used, the commands are in the form of a velocity vector of the TCP in one of the above-mentioned directions.

In operating mode, trajectory interpolators are required for two purposes:

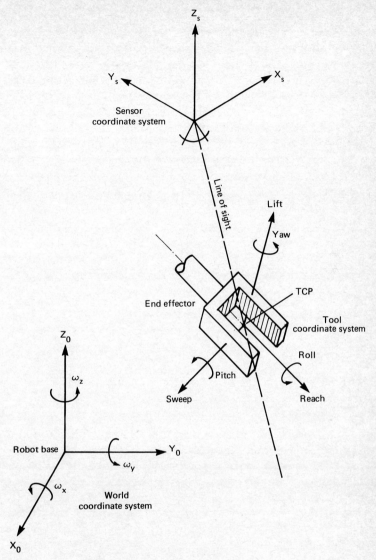

FIG. 5-1 Useful coordinate systems in CP robot.

1. In most present CP robots the interpolator is used to generate motions along straight-line trajectories between programmed absolute end points. These trajectories are useful in some robot applications, such as machine loading, assembly, arc welding, and deburring. The location of the end points can be specified in the task program, or alternatively, can be calculated in real-time during

operation. One example for the latter case is when a television camera is used to identify, in real-time, the location of objects and accordingly to generate the corresponding trajectory. Another example is the palletizing operation. With CP robots it is sufficient to program one point on the pallet, the orientation of the pallet, and the relative position in which the objects should be arranged on the pallet. The location of each absolute position on the pallet and the corresponding trajectories are calculated by the robot computer in real time and transformed by the interpolator to the associated joint values. In this case, it is not necessary to program every single location in the pallet, which is the case when using PTP robots.

2. In several intelligent robots the interpolator is used to generate motions along random trajectories which are continuously calculated in real time. For example, in the deburring operation, the contact force between the workpiece and the robot tool can be measured continuously by a tactile sensor and subsequently used to calculate the desired instantaneous displacements of the tool.

5.1.2 The Generation of Motion Commands

Motion commands in CP robots are generated through a hierarchical structure such as shown in Fig. 5-2. In intelligent robots (see Chap. 8), at the top level is the artificial intelligence (AI) algorithm, which accepts information from the associated sensors as well as from the task program. The AI algorithm calculates in real time a target point, consisting of a desired position and orientation of the end effector, and sends it to the interpolator. In nonintelligent robots, the target point is directly supplied by the task program.

At the intermediate level is the interpolator which has two major tasks: trajectory planning and solving the inverse kinematics. The trajectory-planning algorithm specifies a series of intermediate points along the trajectory between the present TCP location and the desired one, as shown in Fig. 2-6. These points must be within the reach envelope of the manipulator and the motion through them should be subjected to spatial velocity and acceleration constraints.

Next the interpolator applies an inverse kinematics algorithm to solve for the desired axial positions (q) and speeds ($\dot{q}$), which are sent as new set points to the control loops. At the lowest level there is the controller of the individual loops, which controls the motion of the corresponding axes. The three levels of the hierarchy operate at three different sampling intervals, T_b, T_a, and T_t, which are related to the control loops, the interpolator, and the AI algorithm, respectively. In all robots the relationship $T_b \leq T_a \leq T_t$ exists.

One of the design objectives in robotics is to shorten the sampling

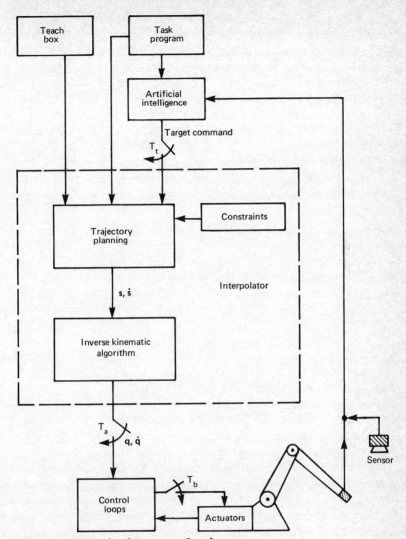

FIG. 5-2   Hierarchical structure of a robot system.

periods as far as possible. Shorter T_b causes smaller follow-up errors and smaller overshoots. Shorter T_a permits smaller sections when dividing the trajectory, which in turn causes improved tracking of the desired paths and better performance when following time-dependent trajectories in intelligent robots. The sampling periods are limited by the execution time of the algorithms. However, the software algorithms of the three hierarchical levels are almost independent, and therefore they can be computed in parallel, i.e., in separate computers. Parallel computation permits operation with the shortest sampling periods possible as well as facilitates the introduction of software modifications at the development stage.

5.1.3 The Trajectory Planning

The trajectory-planning algorithm divides the path between the segment's end points, or between the present point and the target point, into small sections by adding intermediate points along the path. These points are sent to the inverse kinematics algorithm which, in turn, sends axial commands every T_a seconds.

The target point can be defined in two ways:

1. By an *end point* of a trajectory, given by either the task program or by a sensor which can identify the desired final position. This is the most common case which is found in all present (1984) commercial robots, and in many intelligent robots. In this case the target command is given once at the beginning of the trajectory.

2. By a spatial *displacement increment* defined by its direction and magnitude. In this case the target command is continuously changed in a random manner during the interpolation. This is common in adaptive robots in which the sensor is used to guide the robot along a desired trajectory. In this case the sensor is continuously used as a feedback device for the AI level. The most significant difference between this version and the previous one is that here the AI level serves as a controller which is included in an overall control loop. The effect of the interpolator on this control loop is similar to a dead-time element, caused by the computation time of the algorithm.

Target commands must be transformed into motion commands which are sent to the inverse kinematics algorithm in the form of either a position vector or a velocity vector. For a six-axis manipulator, the position command s is a six-component array vector where three components describe the position of the TCP and the other three describe the orientation of the end effector. Likewise, the velocity command $\dot{s}$ is a six-component array vector describing the rate of change of s. Both s and $\dot{s}$ should satisfy the spatial velocity and acceleration constraints.

The velocity command vector is expressed in the form of a displacement increment Δs, according to [see Eq. (2-18)]

$$\Delta s(i) = \dot{s}(i) T_a \qquad (5\text{-}1)$$

where $\dot{s}(i)$ is the value of $\dot{s}$ at $t = iT_a$ (i is an integer), and $\Delta s(i)$ is the required displacement during the following sampling period. Note that

T_a is a constant. The corresponding position command s(i) is calculated by accumulating the displacement increments according to

$$s(i) = s(i - 1) + \Delta s(i) \tag{5-2}$$

where $s(i - 1)$ and $s(i)$ denote the previous and current positions, respectively.

The trajectory-planning algorithm calculates the desired acceleration at the beginning of the trajectory and the deceleration before reaching the end point. The appropriate velocity command is designed in a form similar to Fig. 2-5 with $\dot{s}$ replacing V.

The acceleration stage can be performed as follows: starting with $\Delta s(0) = 0$, the increments are recalculated during each sampling period T_a according to

$$\Delta s(i) = \Delta s(i - 1) + a_1 T_a^2 \tag{5-3}$$

where $\Delta s(i - 1)$ and $\Delta s(i)$ denote the previous and the current displacement commands, and a_1 is the required acceleration along the path. The acceleration stage continues until the desired velocity $\dot{s}$ reaches its maximum value (i.e., maximum Δs). Then Δs remains constant until the deceleration stage, where it is decreased according to Eq. (5-3) with a deceleration $-a_2$ replacing $+a_1$.

When curved trajectories are generated, the centrifugal acceleration constraint is required. This constraint is introduced in the form of a minimum allowable radius of curvature, which corresponds to the allowable acceleration and to the magnitude of $\dot{s}$ (Lewis, 1974). Such curved trajectories are required, for example, when via points are considered.

A via point is one through which the TCP should approximately pass without stopping. The corresponding trajectory is composed of two straight lines connected through a curved path as shown in Fig. 5-3. The radius of curvature of this trajectory, ρ, corresponds to the maxi-

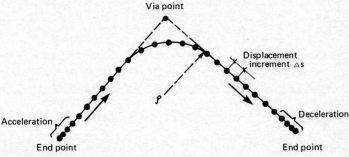

FIG. 5-3   The use of a via point in robot trajectory.

mum permitted acceleration of the TCP. Notice that if an attempt to perform a sharp change in the direction is made, the large acceleration demanded at the TCP cannot be achieved and the resulting motion is undesirable.

5.1.4 Basic Structure of Interpolators

Concerning the mathematical complexity and the number of arithmetic operations involved, the inverse kinematics algorithm is considered the major part of the interpolator. Therefore, the term *interpolator* will refer henceforth to the inverse kinematics algorithm itself.

The inputs to the interpolator are the spatial position (s) and/or speed ($\dot{s}$) vectors which are calculated in the trajectory-planning stage. In the interpolator, these vectors are transformed into the associated axial positions (q) or speeds ($\dot{q}$) of the axes of motion, and subsequently used as reference signals to the control loops. In an n-axis manipulator, q and $\dot{q}$ are n-component array vectors. The transformation, which is based on the solution of the inverse kinematics problem, is performed during each sampling interval T_a.

Interpolators can be classified as absolute or incremental types. In the absolute interpolator, the position s is transformed into the absolute axial positions q. The incremental interpolator transforms the velocity $\dot{s}$ into the corresponding axial speeds $\dot{q}$.

Absolute Interpolators.   With these interpolators, at each sampling period i a new absolute axis position $q(i)$ is calculated, as shown in Fig. 5-4a. The components of $q(i)$ are sent as position references to the individual control loops. It is expected that at time $t = (i + 1)T_a$ the actual position of each axis will reach the corresponding reference position.

If a reference velocity is required, it is calculated from the absolute axis position. The position difference between the present and previous absolute axis positions is the displacement vector Δq given by

$$\Delta q(i) = q(i) - q(i - 1) \tag{5-4}$$

The required axis speed vector is proportional to the displacement vector

$$\dot{q}(i) = \frac{\Delta q(i)}{T_a} \tag{5-5}$$

The reference speed remains constant during the subsequent sampling period, as shown in Fig. 5-4b.

With the absolute interpolator there is no reference position error at the intermediate points. However, between these points there is an

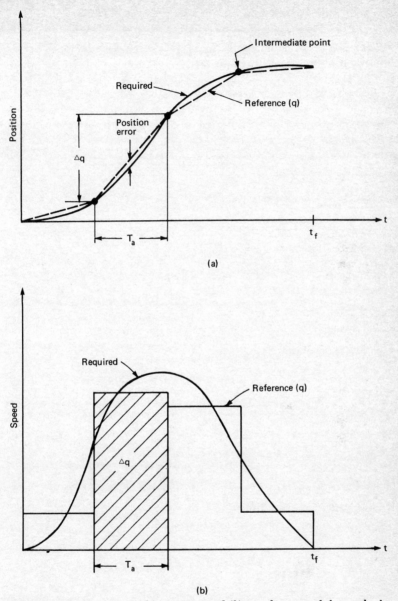

FIG. 5-4 The generation of (*a*) position and (*b*) speed commands by an absolute interpolator.

135

error (see Fig. 5-4a) which depends on the magnitude of $\Delta\mathbf{q}$, and consequently the error is larger for longer T_a or for faster speed, as shown in Fig. 5-4b. Additional position errors are caused by computational errors in the interpolator algorithm, e.g., truncation errors and errors resulting from other mathematical approximations. However, since the calculation of a reference position is not affected by the previous positions, these errors do not accumulate during the interpolation process.

Another computational difficulty in absolute interpolators is due to the fact that the solution of the inverse kinematics problem is not unique. In other words, there are several axial positions that result in the same spatial position (Brady, 1982; Paul, 1981), but only one of them suits the requirement. This situation can happen, for example, when inverse trigonometric functions are used and the resulting angular values are not unique. A simple case which shows that a specific position can be achieved by two different arm configurations will be shown later in Figs. 5-9 and 5-11. The interpolator algorithm must select the desired solution.

Incremental Interpolators.   With this technique, at each sampling period the spatial velocity vector $\dot{\mathbf{s}}(i)$ is transformed by the interpolator into the axis speed vector $\dot{\mathbf{q}}(i)$. This speed is approximated to the staircase shape shown in Fig. 5-5a and is fed as a reference to the control loop. The references are held constant during the interval T_a. To obtain the axis position, a digital integration is performed by approximating the area below the speed curve as a sum of rectangular areas Δq, each of equal base T_a

$$\Delta\mathbf{q}(i) = \dot{\mathbf{q}}(i)T_a \tag{5-6}$$

The digital integration of Eq. (5-6) yields

$$\mathbf{q}(i) = \sum_{j=1}^{i} \Delta\mathbf{q}(j) \tag{5-7}$$

Each component of $\mathbf{q}$ is an axis reference. The dashed line in Fig. 5-5b represents a reference position. The required position can be expressed by

$$\mathbf{q}(t) = \int_{0}^{t} \dot{\mathbf{q}}(\tau)\, d\tau \tag{5-8}$$

and is shown by the solid line in Fig. 5-5b. The difference between the required position and the reference position is the axis position error. The combination of all the axis position errors in space is the trajectory-

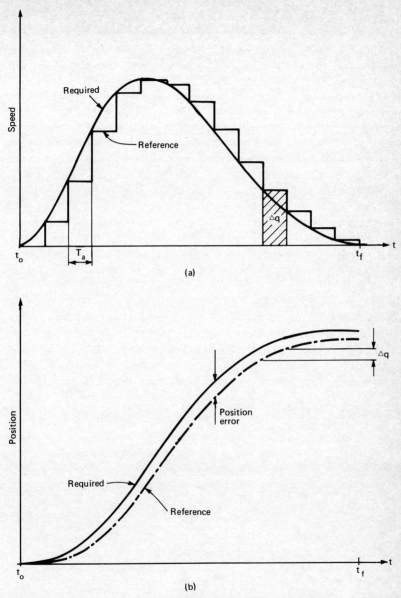

FIG. 5-5 The generation of (*a*) speed and (*b*) position commands by an incremental interpolator.

following error. This error accumulates during the interpolation cycles, resulting in a final position error that cannot be corrected by the incremental interpolator.

Since position errors are accumulated, the incremental interpolator is more sensitive to computational errors (e.g., truncation errors). Note that if the sampling interval is reduced in order to achieve more accurate integration in Eq. (5-7), the number of interpolation cycles is increased, and consequently the computation errors become more significant.

The above-mentioned position errors cause additional trajectory-following errors, as shown in Fig. 5-6. Consider a straight-line trajectory defined by start and end points. The spatial displacement increments Δs that are calculated in the trajectory-planning level represent spatial direction vectors along the trajectory. Assume that after the first interpolation cycle, a position error is generated. In the next interpolation cycle, Δs represents a direction which is parallel to the original trajectory, but it is no longer pointing toward the end point. This procedure is repeated during the following interpolation cycles, causing an increasing position error, as shown in Fig. 5-6.

This error can be corrected in the trajectory-planning level. The correction includes two steps (see Fig. 5-7). First the actual position $s^*(i)$ is calculated from the present axial position $q(i)$ [given in Eq. (5-7)] using direct kinematics. Then $s^*(i)$ is compared with the next spatial position $s(i + 1)$, generating the desired displacement, $\Delta s(i + 1)$, which

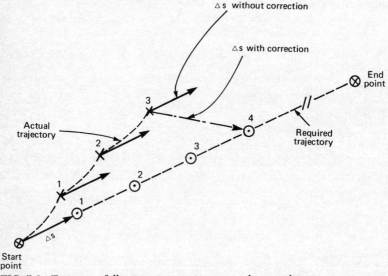

FIG. 5-6 Trajectory following errors in incremental interpolators.

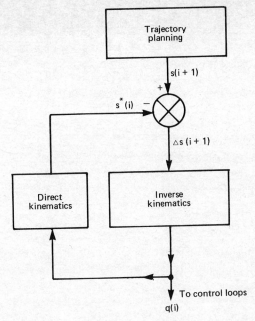

FIG. 5-7 A correction loop for trajectory errors
in robots with incremental interpolators.

points toward the required position on the trajectory (see Fig. 5-6).
This displacement should cause the TCP to converge back to the trajectory. When used, this correction compensated for position errors
caused by the incremental interpolator, and therefore they are no
longer accumulated.

In both absolute and incremental algorithms, the reduction of the
spatial increment Δs should increase the positional accuracy. Since Δs
is smaller with slow traveling speed [see Eq. (5-1)], there is a trade-off
between the speed and the positional accuracy. However, with short
sampling period T_a, precise interpolation accuracy can be achieved
even at high speed.

It should be noted that additional trajectory-following errors are
caused by the position errors in the control loops and by mechanical
inaccuracies.

5.1.5 The Solvability of the Inverse
Kinematics Problem

The direct kinematics problem in robotics is always defined. That
means that the position and orientation of the end effector can always
be computed from a given axis position vector **q**. By contrast, the

inverse kinematics problem is not always solvable, meaning that it is not always possible to achieve any desired position or orientation with a given manipulator. The ability to position and orient the end effector in space depends on the number and the type of axes, and on their kinematic configuration.

As an example, consider the spherical arm shown in Fig. 4-2b. The inverse kinematics of this arm can be solved only for the position of the TCP, but not for its orientation. That is, it is possible to calculate the joint values which bring the TCP to any desired position within the reach envelope of the arm, but not with a desired orientation; and it is impossible to orient the TCP even if the position is not specified. However, if the prismatic point 3 in Fig. 4-2b is replaced by a revolute joint (around axis Z_2), it becomes possible either to orient the TCP or to position it.

When both position and orientation are required, the manipulator should have six independent axes of motion. In all practical six-axis robots, the inverse kinematics problem is usually solvable within the reach volume of the manipulator. There are, however, particular cases in which the inverse kinematics problems is not solvable, for example, when two revolute axes coincide (e.g., axes 4 and 6 of the PUMA); thus they are no longer independent. This phenomenon, which causes degeneration of degrees of freedom, is known as *singularity*. Singular positions are rarely found in most existing four- or five-axis manipulators, but they are most frequently encountered in six-axis manipulators.

Since the singularity is a phenomenon associated with kinematic configuration, it is independent of the technique used to solve the inverse kinematics problem. The various techniques may differ in their ability to identify these situations and recover from them. Whenever the manipulator approaches a singular position, the computational errors at this point may cause large positional errors, or alternatively, disable the coordinate transformation. In general, incremental algorithms are more sensitive to this phenomenon.

While the advantages of the absolute interpolator over the incremental one are obvious, the absolute interpolator is not always feasible because of mathematical complexity in solving the inverse kinematics.

5.2 PARTICULAR SOLUTIONS FOR THE INVERSE KINEMATICS PROBLEM

The derivation of absolute and incremental algorithms for two-axis planar mechanisms is discussed in Sec. 5.2.1. Examples for three-axis arms are given in Sec. 5.2.2 and six-axis manipulators are considered in Sec. 5.2.3.

5.2.1 Two-Axis Planar Mechanisms

Three basic types of mechanisms are discussed: a polar, an articulated, and an indirect drive.

Polar Mechanism.   The axial positions q and $\dot{q}$ of the polar mechanism shown in Fig. 5-8 are

$$q = \begin{bmatrix} \theta \\ r \end{bmatrix} \qquad \dot{q} = \begin{bmatrix} \dot{\theta} \\ \dot{r} \end{bmatrix} \tag{5-9}$$

The spatial (or planar in this case) position and velocity vectors are

$$s = \begin{bmatrix} X \\ Y \end{bmatrix} \qquad \dot{s} = \begin{bmatrix} \dot{X} \\ \dot{Y} \end{bmatrix} \tag{5-10}$$

One approach to solving the inverse kinematics problem is by first deriving the direct kinematics equations and then solving them for the axial variables. Likewise, the direct kinematics equations for Fig. 5-8 are

$$X = r \cos \theta \qquad Y = r \sin \theta \tag{5-11}$$

and the corresponding *absolute algorithm* is

$$\theta = \tan^{-1} \left(\frac{Y}{X} \right) \pm k\pi \qquad r = (X^2 + Y^2)^{1/2} \tag{5-12}$$

where k is either zero or one. If the operating range of the θ axis is bounded within two quadrants in the XY plane, then the solution in Eq. (5-12) is unique (e.g., if θ is expected in the range $-90°$ to $+90°$, then $k = 0$). If, however, the range is not limited, Eq. (5-12) offers two alternative solutions for θ which differ by $180°$. In this case, the undesired solution is easily recognized since it does not satisfy Eq. (5-11). Equa-

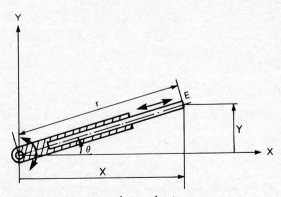

FIG. 5-8 Two-axis polar mechanism.

tion (5-12) is the basis for the absolute interpolator that was discussed in Sec. 5.1.4.

In order to establish an *incremental algorithm*, $\dot{s}$ should be resolved into $\dot{q}$. In this case the solution is based upon the time derivative of Eq. (5-11):

$$\begin{bmatrix} \dot{X} \\ \dot{Y} \end{bmatrix} = \begin{bmatrix} -r \sin \theta & \cos \theta \\ r \cos \theta & \sin \theta \end{bmatrix} \begin{bmatrix} \dot{\theta} \\ \dot{r} \end{bmatrix} \tag{5-13}$$

which can also be presented in a matrix form

$$\dot{s} = J\dot{q} \tag{5-14}$$

where the matrix **J** is called the *Jacobian matrix* or the *influence matrix* of the corresponding mechanism.

The inverse kinematics solution is obtained by solving Eq. (5-14) for $\dot{q}$

$$\dot{q} = J^{-1}\dot{s} \tag{5-15}$$

where the inverse of the Jacobian matrix is given by

$$J^{-1} = \frac{1}{r} \begin{bmatrix} -\sin \theta & \cos \theta \\ r \cos \theta & r \sin \theta \end{bmatrix} \tag{5-16}$$

The solution of Eq. (5-15) is unique.

The corresponding incremental interpolator is obtained by replacing $\dot{q}$ and $\dot{s}$ in Eq. (5-15) by the displacement increments $\Delta q = [\Delta\theta, \Delta r]^T$ and $\Delta s = [\Delta X, \Delta Y]^T$

$$\Delta\theta(i) = \frac{1}{r(i)} [-\Delta X(i) \sin \theta(i) + \Delta Y(i) \cos \theta(i)]$$
$$\Delta r(i) = \Delta X(i) \cos \theta(i) + \Delta Y(i) \sin \theta(i) \tag{5-17}$$

Equations (5-17) are calculated during each sampling interval i, and the resultant values of $\Delta\theta$ and Δr are used as references to the control loops.

In order to correct the accumulated trajectory-following errors, the difference Δs between the required position $s(i)$ and the current actual position, $s^*(i - 1)$ should be generated (see Fig. 5-6). According to Fig. 5-7, and for $s^*(i - 1)$ calculated from Eq. (5-11), $\Delta s(i)$ is given by

$$\Delta X(i) = X(i) - r(i - 1) \cos \theta(i - 1)$$
$$\Delta Y(i) = Y(i) - r(i - 1) \sin \theta(i - 1) \tag{5-18}$$

The sequence of operations during one interpolation cycle is as follows: first $\Delta s(i)$ is calculated from Eq. (5-18) and then it is used in Eq. (5-17) to solve for $\Delta q(i)$. Note that the values of the trigonometric function in Eq. (5-18) are calculated in the previous cycle.

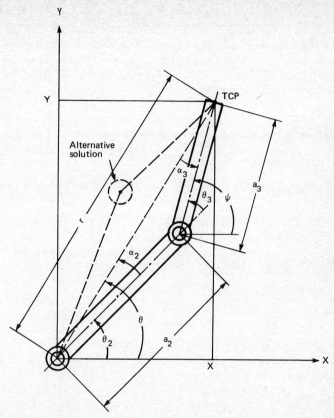

FIG. 5-9 Two-axis planar articulated mechanism; two configurations can bring the TCP to a desired location.

Articulated Mechanism.   The articulated mechanism shown in Fig. 5-9 is similar to that in Fig. 4-2a. The axial position **q** and speed **q̇** are

$$\mathbf{q} = \begin{bmatrix} \theta_2 \\ \psi \end{bmatrix} \qquad \dot{\mathbf{q}} = \begin{bmatrix} \dot{\theta}_2 \\ \dot{\psi} \end{bmatrix} \tag{5-19}$$

where

$$\psi = \theta_2 + \theta_3 \tag{5-20}$$

The spatial vectors s and ṡ are given in Eq. (5-10).

The solution of the direct kinematics problem is given in Eq. (4-1)

$$\begin{aligned} X &= a_2 \cos \theta_2 + a_3 \cos \psi \\ Y &= a_2 \sin \theta_2 + a_3 \sin \psi \end{aligned} \tag{5-21}$$

One approach to find an *absolute algorithm* is by deriving $\mathbf{q}$ from Eq. (5-21). However, it is relatively complicated in this case. An alternative approach is a straightforward geometric approach. The definition of the radius vector of the TCP, r, and the angle θ in Fig. 5-9 are similar to those in Fig. 5-8, and therefore they are calculated from Eq. (5-12). The angles α_2 and α_3 are derived from applying the law of cosines to the triangle formed by the two links and the radius vector

$$\alpha_2 = \pm\cos^{-1}\left(\frac{r^2 + a_2^2 - a_3^2}{2ra_2}\right)$$

$$\alpha_3 = \pm\cos^{-1}\left(\frac{r^2 + a_3^2 - a_2^2}{2ra_3}\right) \tag{5-22}$$

Subsequently, θ_2 and ψ are given by

$$\theta_2 = \theta - \alpha_2 \qquad \psi = \theta + \alpha_3 \tag{5-23}$$

where θ and r are given by Eq. (5-12). Note that the condition

$$|a_2 - a_3| \leqq r \leqq a_2 + a_3 \tag{5-24}$$

should be satisfied. If the angle θ_3 is desired, it can be calculated from Eq. (5-20).

The solution of $\dot{\mathbf{q}}$ is not unique in this case, since Eqs. (5-12) and (5-22) offer two alternative values to each of the angles θ, α_2, and α_3. The consideration for choosing the desired solution for θ is similar to that of the polar case. Additional considerations are required, however, for the two angles α_2 and α_3. From Fig. 5-9 it can be seen that there are two configurations which bring the TCP to its desired location. One configuration is generated when both α_2 and α_3 are positive, and the other when they are negative; the other two possible combinations with unequal signs are rejected. (Unequal signs of α_2 and α_3 mean a mechanical configuration where the two links are disconnected.) Subsequently, it is desired to choose between the two feasible alternative configurations, shown in Fig. 5-9. Since a continuous motion of the mechanism is considered, the sign of α_2 and α_3 is chosen equal to that in the previous axial position. When the magnitude of $(\alpha_2 + \alpha_3)$ approaches $0 \pm k\pi$, then the mechanism approaches its singular position in which

$$r = |a_3 \pm a_2| \tag{5-25}$$

In this case, it may be desired to change the sign of α_2 and α_3 with respect to the previous position, and additional considerations are then required.

The *incremental algorithm* is derived according to Eq. (5-15), where the corresponding Jacobian matrix is obtained from the time derivative of Eq. (5-21)

$$\mathbf{J} = \begin{bmatrix} -a_2 \sin \theta_2 & -a_3 \sin \psi \\ a_2 \cos \theta_2 & a_3 \cos \psi \end{bmatrix} \tag{5-26}$$

Substituting the inverse of $\mathbf{J}$ into Eq. (5-15) and replacing $\dot{\mathbf{q}}$ and $\dot{\mathbf{s}}$ by $\Delta \mathbf{q}$ and $\Delta \mathbf{s}$, respectively, yields the incremental algorithm

$$\begin{bmatrix} \Delta \theta_2 \\ \Delta \psi \end{bmatrix} = \frac{1}{a_2 a_3 \sin(\psi - \theta_2)} \begin{bmatrix} a_3 \cos \psi & a_3 \sin \psi \\ -a_2 \cos \theta_2 & -a_2 \sin \theta_2 \end{bmatrix} \begin{bmatrix} \Delta X \\ \Delta Y \end{bmatrix} \tag{5-27}$$

The spatial displacement $\Delta \mathbf{s}$ is calculated according to Fig. 5-7 as the difference between the required position $s(i)$ and the actual position which is given in Eq. (5-21)

$$\Delta X(i) = X(i) - [a_2 \cos \theta_2(i-1) + a_3 \cos \psi(i-1)] \tag{5-28}$$
$$\Delta Y(i) = Y(i) - [a_2 \sin \theta_2(i-1) + a_3 \sin \psi(i-1)]$$

During each interpolation cycle $\Delta \mathbf{s}$ is calculated from Eq. (5-28) and subsequently it is used in Eq. (5-27) to derive the axial displacement $\Delta \mathbf{q}$.

In the two singular positions described in Eq. (5-25), the following relation exists:

$$\psi = \theta_2 \pm k\pi \qquad k = 0 \text{ or } 1 \tag{5-29}$$

and therefore the denominator of Eq. (5-27) becomes zero and the computation of $\Delta \mathbf{q}$ is disabled. In this case the singularity is related to the degeneration of the mechanism, which now constitutes only one degree of freedom (the two links are on the same line), and both $\dot{\theta}_2$ and $\dot{\psi}$ result in the same spatial direction.

Figure 5-10 demonstrates a singularity problem which might occur at the end of the reach envelope [see Eq. (5-25)]. Assume that the TCP of the manipulator shown in Fig. 5-9 has to move along the straight line segment AC. Note that point C is very nearly at the end of the reach envelope, since the segment OC is unaccessible by the manipulator. Let us divide the path AC into two equal segments, such that $AB = BC$. As is seen from Fig. 5-10b, the motion from A to B requires a small change in the angles ψ and θ, but to produce the equivalent motion from B to C, a significant angular change is needed. This phenomenon is even more emphasized when observing the corresponding joint speeds as shown in Fig. 5-10c. The speed is smooth in the region AB, but drastically changes in the segment BC. Note that the motion in the neighborhood of point C requires theoretically infinite speeds. In the robot computer the corresponding algorithm will try to perform illegal mathematical operations at point C, such as a division by zero, and will fail. This is what is known as a singularity phenomenon in manipulators.

A computational problem exists not only at point C but also in its neighborhood (on the left side of the dashed line in Fig. 5-10c). In this

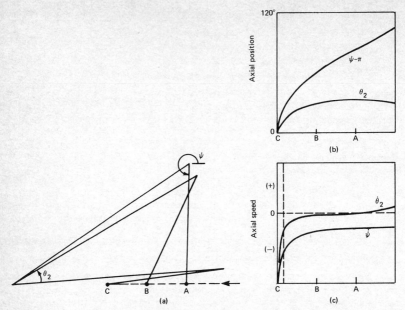

FIG. 5-10 Approaching a singular point at the end of the reach envelope. (*a*) The manipulator; (*b*) the change in the joint angles; (*c*) the change in speed in the neighborhood of the singular point.

region, the speed curves are so sensitive that small computational errors result in enormous speed errors. As a consequence, this region is characterized by unexpected sharp accelerations, unpredictable trajectories, and even position errors at the end points.

Substantial position errors are introduced in this region when employing the incremental interpolator algorithm. As was illustrated in Fig. 5-5, with this algorithm the position calculation is based upon a rectangular approximation of the speed which results in position errors proportional to the change in speed during the iteration time T_a. In the neighborhood of point C the change in speed is large, and the speed itself is inaccurate (since the algorithm includes a division by a small number), which results in substantial position errors. By contrast, the absolute interpolator does not introduce position errors. However, problems associated with sharp accelerations remain when applying this algorithm as well.

Another singularity phenomenon is illustrated in Fig. 5-11, where a motion from point A to point B through C is required. There are two possible manipulator configurations in which the TCP can be located at point A (a_1 and b_1) and two configurations for B (a_2 and b_2). When the TCP passes through point C, any of the two configurations, a_2 or b_2, can

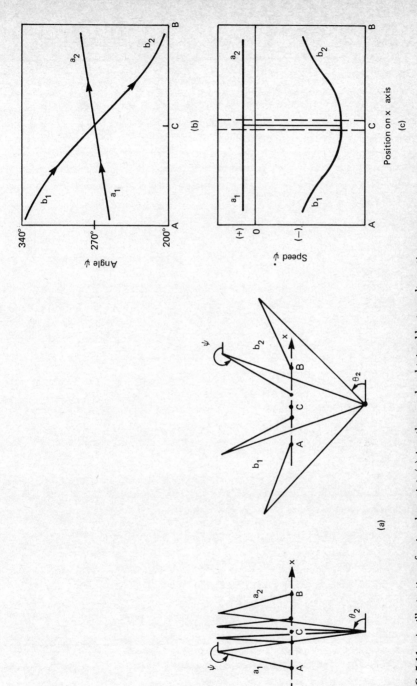

FIG. 5.11 An illustration of a singular point. (*a*) A path *x* can be obtained by two alternative configurations of the manipulator; (*b*) the change in the joint angle ψ; (*c*) the change in speed in the neighborhood of the singular point C.

147

be obtained, and the one selected by the computer depends on the particular algorithm and its computational errors. The basic difference between this case and the previous one is that here the speed does not approach infinity at the singular point C. (The reason is that the algorithm contains an expression of the type x/y; both x and y are zero at C but x approaches zero more rapidly than y.)

Applying the absolute interpolator in this case does not introduce

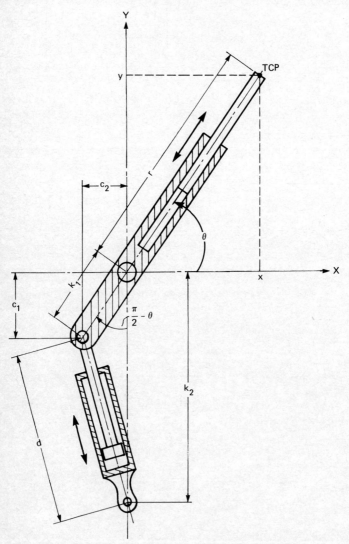

FIG. 5-12   Two-axis manipulator with closed-kinematic chain.

position errors. The selection of the appropriate solution a_2 or b_2 depends on the sign ($+$ or $-$) of α_2 and α_3 in Eq. (5-22). By contrast, the incremental algorithm, which is based on Eq. (5-27), results in position errors in the neighborhood of C. The reason is that both the numerator and the denominator of Eq. (5-27) are very small in this region and truncation errors become dominant. Another consequence of the truncation errors is that either configuration a_2 or b_2 can be obtained when the arm moves through C.

Indirect-Drive Mechanism.   Mechanisms in which the motion of an actuator is transmitted to the joint via subkinematic chains are denoted as indirect-drive mechanisms. In such a mechanism, illustrated in Fig. 5-12, the joint θ is indirectly actuated by the linear actuator d, through a slider-crank mechanism (where d is the slider and k_1 is the crank). In this case, the spatial vectors (s or ṡ) should be resolved into either of the axis vectors

$$\mathbf{q} = \begin{bmatrix} d \\ r \end{bmatrix} \quad \text{or} \quad \dot{\mathbf{q}} = \begin{bmatrix} \dot{d} \\ \dot{r} \end{bmatrix} \tag{5-30}$$

The solution of the inverse kinematics problem for this structure can be based on the solution derived for r and θ in Fig. 5-8. For an absolute algorithm, the position of d is calculated from θ, applying the law of cosines to the triangle (k_1, k_2, d) in Fig. 5-12

$$d = (k_1^2 + k_2^2 - 2k_1k_2 \sin \theta)^{1/2} \tag{5-31}$$

where θ is given in Eq. (5-12).

An alternative approach is to derive the displacement d directly from s. Since the angles of the triangles (k_1, c_2, c_1) and (r, x, y) in Fig. 5-12 are similar, the parameters c_1 and c_2 are given by

$$c_1 = \frac{k_1}{r} y \qquad c_2 = \frac{k_1}{r} x \tag{5-32}$$

with r given in Eq. (5-12), x and y correspond to s, and k_1 is a constant. Subsequently, d is given by

$$d = [(k_2 - c_1)^2 + c_2^2]^{1/2} \tag{5-33}$$

where k_2 is a constant.

If an incremental solution is desired, the relationship between $\dot{d}$ and $\dot{\theta}$ could be derived. However, since the number of arithmetic operations required for the above relationship is significantly larger than the number needed with Eq. (5-33), the incremental solution is inefficient and is thus omitted from this text.

5.2.2 Example of Three-Axis Spherical Mechanism

The procedure demonstrated for two-axis mechanisms can be similarly carried out for three-axis robot arms. Consider, for example, the spherical arm in Fig. 4-2b, for which the following direct kinematics solution is given in Eq. (4-3)

$$X = d_3 \cos \theta_1 \cos \theta_2$$

$$Y = d_3 \sin \theta_1 \cos \theta_2 \tag{5-34}$$

$$Z = d_3 \sin \theta_2$$

Desiring an absolute algorithm, it is complicated to solve Eq. (5-34) for **q**, where

$$\mathbf{q} = \begin{bmatrix} \theta_1 \\ \theta_2 \\ d_3 \end{bmatrix} \tag{5-35}$$

Alternatively, **q** can be derived in two steps according to Fig. 5-13. First, the projection of d_3 on the XY plane, denoted by r, and the angle θ_1 are calculated according to Eq. (5-12), with θ_1 replacing θ. Subsequently, d_3 and θ_2 are calculated in the rZ plane according to Eq. (5-12), yielding

$$\theta_1 = \tan^{-1}\left(\frac{Y}{X}\right) \qquad r = (X^2 + Y^2)^{1/2} \tag{5-36}$$

and

$$\theta_2 = \tan^{-1}\left(\frac{Z}{r}\right) \qquad d_3 = (r^2 + Z^2)^{1/2} \tag{5-37}$$

Note that the solution of the spherical mechanism is presented by consecutive solutions of two polar mechanisms: in the XY plane and then in the rZ plane.

This approach can be implemented to generate an incremental algorithm as well. In this case, the incremental algorithms of the above polar systems [which are in the form of Eq. (5-17)] are consequently calculated: first $\dot{\theta}_1$ and $\dot{r}$ are calculated in the XY plane, and then $\dot{\theta}_2$ and $\dot{d}_3$ are calculated in the rZ plane.

Note that it is possible to apply the above approach to other mechanisms as well. For example, a three-axis articulated mechanism can be represented by two consecutive planar mechanisms, that is, a polar mechanism in the XY plane and an articulated two-axis mechanism (shown in Fig. 5-9) in the rZ plane. In this case the calculation of Eq.

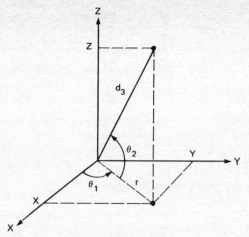

FIG. 5-13 Two consecutive polar systems representing spherical coordinates.

(5-17) in the XY plane is followed by the calculation of Eq. (5-27) in the rZ plane.

5.2.3 Specific Solutions for Six-Axis Manipulators

In six-axis manipulators it is much more complicated to solve the inverse kinematics using straightforward geometry. One reason is the necessity of relating the desired orientation of the end effector to the corresponding axial positions.

Two general techniques for obtaining incremental and absolute algorithms are described later in Secs. 5.3 and 5.4, respectively. In addition, there exists a specific approach (Lee & Ziegler, 1983; Paul, 1981) that permits solution for the axial positions from the direct kinematics solution (with the addition of particular geometric considerations). Therefore, this approach is not general and requires the establishment of a mathematical solution for every individual robot manipulator.

For example, consider a six-axis manipulator for which the displacement matrix of the hand (or the TCP), $\mathbf{T}_0^h$, have been explicitly derived

$$\mathbf{T}_0^h = \left[\begin{array}{c|c} \mathbf{C}_0^h & \mathbf{d}_0^h \\ \hline 0 \quad 0 \quad 0 & 1 \end{array}\right] \tag{5-38}$$

where $\mathbf{C}_0^h$ and $\mathbf{d}_0^h$ are the orientation matrix and the translation vector of the TCP relative to the WCS O_0, respectively. The elements of both $\mathbf{C}_0^h$ and $\mathbf{d}_0^h$ are expressed as a function of the axes' positions $\mathbf{q}$. For a desired spatial position (X^*, Y^*, Z^*) there exists

$$\mathbf{d}_0^h = \begin{bmatrix} X^* \\ Y^* \\ Z^* \end{bmatrix} \tag{5-39}$$

where X^*, Y^*, and Z^* are numerical values. In addition, the desired orientation of the end effector can be used to calculate the numerical values of the nine elements of the desired orientation matrix $\mathbf{C}^*$, for which the following set of equations exists:

$$\mathbf{C}_0^h = \mathbf{C}^* \tag{5-40}$$

Equations (5-39) and (5-40) form a set of 12 equations that should be solved for the six independent variables (i.e., the six elements of $\mathbf{q}$). Since an orientation is defined by three independent parameters, only three independent equations can be derived from the nine equations of Eq. (5-40).† These three equations combined with the three equations of Eq. (5-39) provide the required set of six equations. Since these equations are nonlinear, they can be solved simultaneously only through iterations, or alternatively, the individual elements of $\mathbf{q}$ can be sequentially solved one by one. This latter procedure must be assisted by geometric considerations. For example, it is sometimes possible to define geometrically one of the elements of $\mathbf{q}$ (Lee & Ziegler, 1983). Subsequently, this element, when substituted in several of the six equations, enables the elimination of another variable. This procedure is carried on until all variables are determined.

Note that the complexity of this approach greatly depends on the complexity of the mechanism under consideration. In general this approach suits manipulators which contain geometric simplifications (e.g., three axes of the wrist intersecting in one point or three parallel revolute axes). It should be mentioned that this solution does not suit manipulators which consist of four or five axes, since these manipulators are incapable of generating every desired orientation, and therefore Eq. (5-40) cannot be used.

5.3 RESOLVED MOTION RATE CONTROL METHOD

A general method for deriving an incremental interpolation algorithm was presented by Whitney in 1972 (Whitney, 1972). The method, called the *resolved motion rate control method*, refers to resolving the

---

†There are techniques to identify a set of three independent orientational equations (Mancini et al., 1983; Paul, 1979).

desired motion of the hand into the corresponding axis speeds. This method suits mainly a direct-drive chain manipulator with six axes of motion.

5.3.1 Resolved Rate Strategy

The desired motion of the end effector is described by two vectors $\dot{x}$ and ω

$$\dot{x} = \begin{bmatrix} \dot{X} \\ \dot{Y} \\ \dot{Z} \end{bmatrix} \qquad \omega = \begin{bmatrix} \omega_x \\ \omega_y \\ \omega_z \end{bmatrix} \qquad (5\text{-}41)$$

The vector $\dot{x}$ is the velocity of the origin of the TCS attached to the hand, and the vector ω is the rotation rate of this frame (see Fig. 5-1). Both vectors can be expressed in any of the spatial coordinate systems, WCS, TCS, or SCS, shown in Fig. 5-1. If, for example, these vectors are given in the TCS, the elements of $\dot{x}$ describe the velocity of the TCP along the reach, lift, and sweep axes, and the elements of ω give the rotation rate of the end effector about these axes.

A six-component array vector $\dot{s}$, denoted as the command vector, is given by

$$\dot{s} = \begin{bmatrix} \dot{x} \\ \omega \end{bmatrix} \qquad (5\text{-}42)$$

where $\dot{x}$ and ω are given in Eq. (5-41). The corresponding axis speeds are arranged in a form of a six-component array vector $\dot{q}$.

$$\dot{q} = \begin{bmatrix} q_1 \\ q_2 \\ \vdots \\ q_6 \end{bmatrix} \qquad (5\text{-}43)$$

The two vectors $\dot{s}$ and $\dot{q}$ can be related in a matrix form

$$\dot{s} = J(q)\dot{q} \qquad (5\text{-}44)$$

where $J(q)$ is a 6×6 matrix whose elements are dependent on the axial position vector q and is called the Jacobian matrix or the influence matrix.

The objective of resolving the command vector $\dot{s}$ into the axial speeds $\dot{q}$ may be achieved by inverting the matrix $J(q)$, which yields

$$\dot{q} = [J(q)]^{-1}\dot{s} \qquad (5\text{-}45)$$

The resolved motion rate control method presents a general technique to obtain the Jacobian matrix in Eq. (5-44) and suggests several alternative solutions for $\dot{q}$ [e.g., Eq. (5-45)].

Note that Eq. (5-44) can be obtained for manipulators having any number of axes. However, for an n-axis manipulator, $J(q)$ is a $6 \times n$ matrix and cannot be inverted. Therefore, Eq. (5-45) is solvable only for six-axis manipulators.

5.3.2 The Jacobian Matrix for Positioning

Consider a manipulator having only three axes (q_1, q_2, and q_3) with which it is possible to move the hand (or TCP) along a predetermined trajectory in space. For this case, the axial speed vector $\dot{q}$ and the spatial command vector $\dot{s}$ are

$$\dot{q} = [\dot{q}_1, \dot{q}_2, \dot{q}_3]^T \tag{5-46}$$

and

$$\dot{s} = \dot{x} = [\dot{X}, \dot{Y}, \dot{Z}]^T \tag{5-47}$$

where the velocity vector $\dot{x}$ is expressed in the WCS.

By using the inverse kinematics technique it is possible to relate the position of the TCP to the individual axial positions, namely,

$$X = f_1(q_1, q_2, q_3)$$

$$Y = f_2(q_1, q_2, q_3) \tag{5-48}$$

$$Z = f_3(q_1, q_2, q_3)$$

where f_1, f_2, and f_3 are, for example, the three components of the translation vector d_0^3 in T_0^3. The components of $\dot{x}$ are obtained by differentiating Eq. (5-48)

$$\dot{X} = \frac{\partial f_1}{\partial q_1}\,\dot{q}_1 + \frac{\partial f_1}{\partial q_2}\,\dot{q}_2 + \frac{\partial f_1}{\partial q_3}\,\dot{q}_3$$

$$\dot{Y} = \frac{\partial f_2}{\partial q_1}\,\dot{q}_1 + \frac{\partial f_2}{\partial q_2}\,\dot{q}_2 + \frac{\partial f_2}{\partial q_3}\,\dot{q}_3 \tag{5-49}$$

$$\dot{Z} = \frac{\partial f_3}{\partial q_1}\,\dot{q}_1 + \frac{\partial f_3}{\partial q_2}\,\dot{q}_2 + \frac{\partial f_3}{\partial q_3}\,\dot{q}_3$$

or in a matrix form

$$\dot{x} = J(q)\dot{q} \tag{5-50}$$

The Jacobian matrix $\mathbf{J(q)}$ is defined as

$$\mathbf{J(q)} \stackrel{\Delta}{=} \begin{bmatrix} \dfrac{\partial f_1}{\partial q_1} & \dfrac{\partial f_1}{\partial q_2} & \dfrac{\partial f_1}{\partial q_3} \\[2mm] \dfrac{\partial f_2}{\partial q_1} & \dfrac{\partial f_2}{\partial q_2} & \dfrac{\partial f_2}{\partial q_3} \\[2mm] \dfrac{\partial f_3}{\partial q_1} & \dfrac{\partial f_3}{\partial q_2} & \dfrac{\partial f_3}{\partial q_3} \end{bmatrix} \tag{5-51}$$

Each element of $\mathbf{J(q)}$ depends on $\mathbf{q}$ and is defined by

$$J_{ij} = \frac{\partial f_i}{\partial q_j} \tag{5-52}$$

Notice that $i = 1$, 2, and 3 calls for $\dot{X}$, $\dot{Y}$, and $\dot{Z}$, respectively. An equivalent expression for Eq. (5-52) is as follows:

J_{ij} = the ith component of $\dot{x}$ per unit q_j when
all other $\dot{q}_k = 0$ for $k \neq j$.

Each column $\mathbf{J}_j$ results from the contribution of $\dot{q}_j$ to $\dot{x}$ per unit $\dot{q}_j$.

The establishment of $\mathbf{J(q)}$ according to its definition is demonstrated on the manipulator shown in Fig. 4-2b, for which the velocity of the TCP is given in Eq. (4-4) of Chap. 4. Arranging Eq. (4-4) to the form of Eq. (5-49) yields the Jacobian matrix for this case

$$\mathbf{J(q)} = \begin{bmatrix} -d_3 \sin \theta_1 \cos \theta_2 & -d_3 \cos \theta_1 \sin \theta_2 & \cos \theta_1 \cos \theta_2 \\ d_3 \cos \theta_1 \cos \theta_2 & -d_3 \sin \theta_1 \sin \theta_2 & \sin \theta_1 \cos \theta_2 \\ 0 & d_3 \cos \theta_2 & \sin \theta_2 \end{bmatrix} \tag{5-53}$$

where

$$\dot{\mathbf{q}} = [\dot{\theta}_1, \dot{\theta}_2, \dot{d}_3]^\mathrm{T} \tag{5-54}$$

It can be shown that in this case the inverse of $\mathbf{J(q)}$ is

$$\mathbf{J(q)}^{-1} = \begin{bmatrix} -\dfrac{\sin \theta_1}{d_3 \cos \theta_2} & \dfrac{\cos \theta_1}{d_3 \cos \theta_2} & 0 \\[3mm] -\dfrac{\cos \theta_1 \sin \theta_2}{d_3} & -\dfrac{\sin \theta_1 \sin \theta_2}{d_3} & \dfrac{\cos \theta_2}{d_3} \\[3mm] \cos \theta_1 \cos \theta_2 & \sin \theta_1 \cos \theta_2 & \sin \theta_2 \end{bmatrix} \tag{5-55}$$

An incremental algorithm, which is equivalent to that of Sec. 5.2.2, is obtained by substituting $\mathbf{J}^{-1}$ from Eq. (5-55) into Eq. (5-45).

Deriving $J(q)$ from Eq. (5-51) and establishing its explicit inverse suits a three-axis mechanism that can be used for positioning. However, for six-axis manipulators this technique is impractical because

1. In order to derive the three lower rows in $J(q)$, which correspond to ω, three equations that describe the orientation rate of change should be established. The derivation of these equations from the orientation matrix C_0^6 or from the quaternion q is relatively complicated, and an alternative technique is required.

2. It is almost impossible to derive the inverse of the Jacobian matrix explicitly, and therefore numerical techniques should be used to solve for $\dot{q}$ in Eq. (5-44).

The next section describes a vector cross products technique which is used to establish the Jacobian matrix for six-axis manipulators.

5.3.3 The Jacobian Matrix for Positioning and Orienting

Consider a manipulator with six axes (q_i, $i = 1, \ldots, 6$) with which it is possible to command the position and orientation of the end effector. In this case the six-element command vector $\dot{s}$ includes the rotation rate vector ω in addition to the velocity vector $\dot{x}$, as shown in Eq. (5-42). Since the angular velocities (ω_x, ω_y, ω_z) can be also expressed as a function of q and $\dot{q}$ similar to Eq. (5-49), the elements of the Jacobian matrix $J(q)$ are defined similar to Eq. (5-52)

$$J_{ij} = \frac{\partial f_i}{\partial q_j} \qquad i \text{ and } j = 1, 2, \ldots, 6 \qquad (5\text{-}56)$$

where $i = 1$, 2, and 3 corresponds to $\dot{X}$, $\dot{Y}$, and $\dot{Z}$, and $i = 4$, 5, and 6 corresponds to ω_x, ω_y, and ω_z, respectively. An equivalent expression for J_{ij} can be as follows:

$$J_{ij} = \text{the } i\text{th component of } \dot{s} \text{ per unit } \dot{q}_j$$
$$\text{when all } \dot{q}_k = 0 \text{ for } k \neq j$$

Each column vector J_j is the contribution of $\dot{q}_j$ to $\dot{s}$ per unit $\dot{q}_j$.

Calculating J by Vector Cross Products. The cross products technique is based upon the calculation of the contribution of each individual axis speed to the command vector s.

The contribution of $\dot{q}_j$ to the command vector $\dot{s}$ is illustrated in Fig. 5-14, where the jth joint, the WCS, and the TCP are shown. The unit vector $\hat{u}_j$ is pointing to the rotation axis of $\dot{q}_j$, which coincides with the

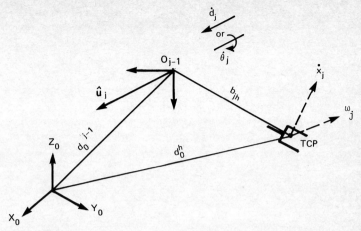

FIG. 5-14   The contribution of the component $\dot{q}_j$ to the command vector $\dot{s}$.

Z_{j-1} axis of frame O_{j-1}. The vector $\mathbf{b}_{jh}$ describes the distance between joint j (the origin of frame O_{j-1}) and the TCP. All vectors are expressed in the WCS (i.e., O_0). The vector $\dot{\mathbf{x}}_j$, which is the contribution of $\dot{q}_j$ to $\dot{\mathbf{x}}$ is given by

$$\dot{\mathbf{x}}_j = (\hat{\mathbf{u}}_j \times \mathbf{b}_{jh})\dot{\theta}_j \qquad \text{for a revolute joint } \theta \tag{5-57}$$

$$\dot{\mathbf{x}}_j = \hat{\mathbf{u}}_j d_j \qquad \text{for a prismatic joint } d$$

Likewise, the vector $\boldsymbol{\omega}_j$, which is the contribution of $\dot{q}_j$ to $\boldsymbol{\omega}$, is given by (see Fig. 5-14):

$$\boldsymbol{\omega}_j = \hat{\mathbf{u}}_j \theta_j \qquad \text{for a revolute joint } \theta \tag{5-58}$$

$$\boldsymbol{\omega}_j = 0 \qquad \text{for a prismatic joint}$$

Subsequently, the column vectors of $\mathbf{J}(\mathbf{q})$ are obtained according to their definition

$$\mathbf{J}_j = \begin{bmatrix} \hat{\mathbf{u}}_j \times \mathbf{b}_{jh} \\ \hat{\mathbf{u}}_j \end{bmatrix} \qquad \text{for a revolute joint} \tag{5-59}$$

$$\mathbf{J}_j = \begin{bmatrix} \hat{\mathbf{u}}_j \\ 0 \end{bmatrix} \qquad \text{for a prismatic joint} \tag{5-60}$$

where the upper and lower portion of $\mathbf{J}_j$ correspond to the contribution of $\dot{q}_j$ to $\dot{\mathbf{x}}$ and $\boldsymbol{\omega}$, respectively. The Jacobian matrix is given by

$$\mathbf{J}(\mathbf{q}) = [\mathbf{J}_1 \quad \mathbf{J}_2 \quad \cdots \quad \mathbf{J}_6] \tag{5-61}$$

The two vectors $\mathbf{b}_{jh}$ and $\hat{\mathbf{u}}_j$ can be calculated from the homogeneous displacement matrices that are derived for the direct kinematics solution (see Chap. 4). The vector $\hat{\mathbf{u}}_j$ is obtained by expressing a unit vector along the Z_{j-1} axis in the O_0 frame (see Fig. 5-14)

$$\hat{\mathbf{u}}_j = \mathbf{C}_0^{j-1} \begin{bmatrix} 0 \\ 0 \\ 1 \end{bmatrix} \tag{5-62}$$

where $\mathbf{C}^{j-1}$ is the orientation matrix of frame O_{j-1} relative to O_0.

The vector $\mathbf{b}_{jh}$ is obtained by subtracting the position of frame O_{j-1} from the position of the TCP (frame O_h)

$$\mathbf{b}_{jh} = \mathbf{d}_0^h - \mathbf{d}_0^{j-1} \tag{5-63}$$

where $\mathbf{d}_0^h$ and $\mathbf{d}_0^{j-1}$ are the translation vectors of $\mathbf{T}_0^h$ and $\mathbf{T}_0^{j-1}$, respectively ($\mathbf{T}_0^h$ denotes the displacement matrix of the TCP relative to frame O_0).

Note that the calculation of $\mathbf{J(q)}$, according to Eqs. (5-59) to (5-63) cannot be practically performed explicitly, thus a numerical computation in the computer is required.

5.3.4 Motions Defined in Other Coordinate Systems

So far, the command vector $\hat{\mathbf{s}}$ was defined in the WCS. If, however, $\hat{\mathbf{s}}$ is defined in other coordinate systems (see Fig. 5-1), then the usage of the Jacobian matrix in Eq. (5-44) requires several modifications.

Motions in the Tool Coordinate System. The command vector $\hat{\mathbf{s}}_h$ expressed in the TCS can be transformed to the corresponding command in the WCS, $\hat{\mathbf{s}}_w$

$$\hat{\mathbf{s}}_w = \mathbf{C}_0^h \hat{\mathbf{s}}_h \tag{5-64}$$

where $\mathbf{C}_0^h$ is the orientation matrix of frame O_h (attached to the hand) with respect to frame O_0. The vector $\hat{\mathbf{s}}_w$ can now be used in Eq. (5-44) with the same Jacobian matrix $\mathbf{J(q)}$ which was derived for the WCS

$$\hat{\mathbf{s}}_w = \mathbf{C}_0^h \hat{\mathbf{s}}_h = \mathbf{J(q)}\dot{\mathbf{q}} \tag{5-65}$$

where $\mathbf{C}_0^h$ multiplies both $\dot{\mathbf{x}}_h$ and ω_h in $\hat{\mathbf{s}}_h$.

An alternative approach suggests the modification of $\mathbf{J(q)}$ rather than of $\hat{\mathbf{s}}$. Multiplying both sides of Eq. (5-65) by the inverse of $\mathbf{C}_0^h$ and recalling that

$$[\mathbf{C}_0^h]^{-1} = [\mathbf{C}_0^h]^T = \mathbf{C}_h^0 \tag{5-66}$$

yields

$$\hat{\mathbf{s}}_h = \mathbf{C}_h^0 \mathbf{J(q)}\dot{\mathbf{q}} \tag{5-67}$$

Now $\dot{s}_h$ remains unchanged while $J(q)$ is multiplied by C_h^0, i.e., each of the vectors $\hat{u}_j \times b_{jh}$ and $\hat{u}_j$ are multiplied by C_h^0.

Motions in Sensors Coordinate System.   If a sensor is mounted on the end effector, then the command vector $\dot{s}$ is directly given in the TCS and interpreted as $\dot{s}_h$. For sensors mounted on one of the manipulator links (e.g., link k) the procedure in Eq. (5-67) is performed, with C_k^0 substituted for C_h^0. If the sensor is fixed in space (e.g., a television camera mounted above the working volume) then another coordinate system, O_s, is attached to it (see Fig. 5-1) and the orientation matrix C_0^s, between the SCS O_s and the O_0 frame, replaces C_0^h in Eq. (5-67).

Another example concerns a motion along a line of sight from an "eye" to the end effector. The line-of-sight vector from the origin of the O_s to the O_h frame is denoted by b_{sh}, and is given by

$$b_{sh} = d_0^h - d_0^s \qquad (5\text{-}68)$$

where d_0^h and d_0^s are the translation vectors of O_6 and O_s with respect to the O_0 frame.

The motion along the line of sight is given in the WCS by Eq. (5-44) with $\dot{x}$ substituting for $\dot{s}$ and given by

$$\dot{x} = \hat{b}_s V \qquad (5\text{-}69)$$

where $\hat{b}_s$ is a unit vector in the direction of b_s, and V is a scalar proportional to the speed of motion. The orientation rate portion of $\dot{s}$ (i.e., w) is zero for a constant spatial orientation.

5.3.5 An Interpolator Based on Resolved Rate Technique

An incremental interpolator (see Sec. 5.1.4), based on the resolved motion rate control method, is described in Fig. 5-15. The main loop of Fig. 5-15 describes the sequence through which the axial speed vector $\dot{q}$ is calculated, and the subloop corresponds to the correction loop shown in Fig. 5-7. At each sampling period T_a, the current axial positions $q(i)$ are used to calculate the corresponding displacement transformation matrices T_{j-1}^j, which in turn are multiplied to generate the displacement transformation matrices of each link relative to the O_0 frame (T_0^i). Subsequently, the Jacobian matrix $J(q)$ is calculated by vector cross products, using $\hat{u}_j$ and b_{jh} from Eqs. (5-62) and (5-63). With the command vector $\dot{s}$, calculated in the trajectory-planning level, Eq. (5-44) is solved for $\dot{q}(i)$, which is the axial speed vector at time $t = iT_a$.

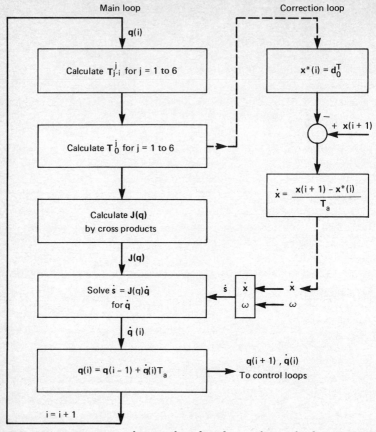

FIG. 5-15 An incremental interpolator based upon the resolved motion rate control method.

The corresponding axial position vector $q(i)$ is calculated according to (see Eq. 5-5)

$$q(i) = q(i - 1) + \dot{q}(i)T_a \tag{5-70}$$

and is used for the next interpolation iteration. Both $q(i)$ and $q(i - 1)$ can be used as references to the control loops.

If only the main loop of Fig. 5-15 is included in the interpolation algorithm, follow-up trajectory errors are generated, as shown in Fig. 5-6. The correction, presented in Sec. 5.1.4 (see also Fig. 5-7), is based on the calculation of $\dot{x}$ from

$$\dot{x} = \frac{x(i + 1) - x^*(i)}{T_a} \tag{5-71}$$

where $x(i + 1)$ is the next position along the trajectory and $x^*(i)$ is the actual position associated with the current value of $q(i)$; that is, $x^*(i)$ is the translation vector d_0^h in T_0^h calculated for $q(i)$. Since the matrix T_0^h is already calculated in the main loop, the correction loop of Fig. 5-15 requires only few additional arithmetic operations.

Computational Features. The main computational difficulty of the above algorithm is to obtain the vector $\dot{q}$ from Eq. (5-44). Since an explicit inversion of the Jacobian matrix [i.e., to the form in Eq. (5-45)] is impractical, there exist several alternative approaches:

1. The Jacobian matrix can be numerically inverted during each interpolation cycle, using matrix inversion routines. This approach requires a relatively large number of arithmetic operations, and consequently large sampling periods T_a (Taylor, 1979).

2. Viewing the reduction of arithmetic operations, Whitney (1972) suggests precomputing several inverse Jacobian matrices along the programmed trajectory. During each computation cycle, each element of the corresponding inverse Jacobian matrix is computed by interpolating between the elements of two precomputed matrices. This approach is less accurate than others and is improper for intelligent robots where trajectories are not known in advance.

3. Another approach (Shoham, 1982) suggests applying numerical techniques that are used to solve a set of equations (e.g., Gauss elimination) in order to solve for $\dot{q}$ in Eq. (5-44). The number of arithmetic operations required in this approach is fewer than in the first one.

All these techniques are sensitive to computational errors, mainly due to the coupling between the various axis parameters (this is well known in numerical analysis of multivariable systems). That is, a small computational error, associated with one of the joints, can result in large errors in the other joints. This is more pronounced in the neighborhood of a singular position. Here, high speeds required from several axes (see Fig. 5-10) result in small numerical values in the corresponding column elements of the Jacobian matrix relative to the other columns. The resulting coupling computationl errors, in this case, are very significant.

In the special case where the manipulator is exactly in a singular position, two (or more) joints can provide only one degree of freedom (i.e., a degeneration of degrees of freedom). Consequently, the corresponding columns of the Jacobian matrix are linearly dependent, and therefore the inversion of $\mathbf{J(q)}$ is not possible. Note that near singularity points, several columns of $\mathbf{J(q)}$ are almost linearly dependent, and the algorithm becomes very sensitive to truncation errors.

5.4 SOLVING THE INVERSE KINEMATICS PROBLEM USING ROTATION VECTORS†

In this section a solution for the inverse kinematics problem is obtained utilizing the rotation vector method for coordinate transformation. The basic equations of using quaternions or rotation vectors to perform coordinate transformation were introduced in Sec. 4.6. Using these equations, the position of the TCP and the tool orientation can be expressed in joint variables $(\mathbf{q})$. In the solution of the inverse problem, the joint variable values are calculated in order to achieve a given position of TCP and a given orientation of the tool.

As was previously explained (Sec. 5.2), there is no general absolute solution of the inverse kinematics problem. Nevertheless, there is an absolute solution for the inverse kinematics problem for almost all existing robots. The reason is that the existing robots maintain some geometric relationships among their axes of motion, such as parallelism or perpendicularity of axes, or axes which intersect in one point. In the following sections two examples demonstrating an inverse kinematics solution of a six-degrees-of-freedom manipulator by using rotation vectors are presented. In these examples the three rotating axes of the wrist are intersecting at one point. This leads to a relatively simple solution, as the motion in these cases can be divided into two parts: position and orientation. The first three degrees of freedom are manipulated to achieve the required wrist intersecting point (i.e., position), and the wrist axes provide the tool's orientation. Note that in this case the motion of the wrist axes drives the TCP on a spherical surface. There are some robots having this kinematic structure, such as the Unimation's PUMA 600 (Fig. 7-1), Cincinnati Milacron's T^3R^3 (Fig. 6-11), Renault's Vertical 80, and GCA's XR6 (Fig. 5-16).

5.4.1 The Inverse Kinematics for a Cartesian Robot

A six-degrees-of-freedom cartesian robot with rotation vectors attached to its wrist axes is shown in Fig. 5-17. There are no rotation vectors in

---

†This section was contributed by M. Shoham.

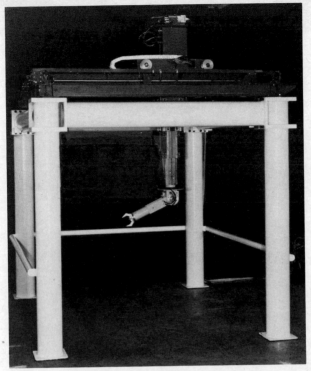

FIG. 5-16 A commercial cartesian robot with the three rotary axes of the wrist intersecting in one point. (GCA.)

the first three axes of motion, as they are linear and permit only parallel motion but cannot change the orientation of the tool. The tool's orientation and position (i.e., the TCP) are given, as well as the tool vector **v**. The tool vector is between O' (the intersection point of the wrist axes) and TCP, and its direction is from O' to TCP. The tool vector is denoted by **v** when the TCP is at the initial position and by **v′** after moving the TCP to a new location.

The joint variables can be derived as follows:

1. Determine the required position of point O' from the initial tool vector **v**, and from the required position and orientation of the tool.

2. Calculate the position of the first three degrees of freedom required to bring the point O' to the new location.

3. Calculate the three wrist angles required to orient the tool in the desired orientation.

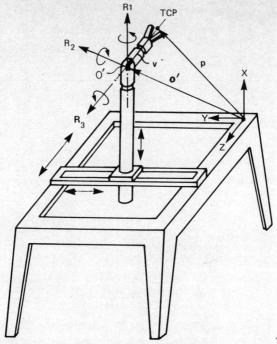

FIG. 5-17   A six-degrees-of-freedom cartesian robot.

Let us consider the following desired position and orientation of the tool:

$$\mathbf{p} = \begin{bmatrix} p_x \\ p_y \\ p_z \end{bmatrix} \qquad \mathbf{R} = \begin{bmatrix} R_x \\ R_y \\ R_z \end{bmatrix} \qquad (5\text{-}72)$$

where $\mathbf{p}$ describes the desired position of TCP and $\mathbf{R}$ denotes the required tool orientation given in a rotation vector form. (Note that in the previous examples in this chapter the required orientation was given by the nine elements of the orientation matrix.) Assuming that the vector $\mathbf{v}$ between O' and TCP is known and originally equals $\mathbf{v} = [v_x, v_y, v_z]^T$. The location of point O' in the base coordinate system can be derived according to the equation

$$\mathbf{o}' = \mathbf{p} - \mathbf{v}' \qquad (5\text{-}73)$$

The vector $\mathbf{v}'$ at its new location is calculated by using vector transformation according to Eq. (4-109)

$$\mathbf{v}' = \mathbf{v} + \frac{2\mathbf{R} \times (\mathbf{v} + \mathbf{R} \times \mathbf{v})}{1 + f^2} \qquad (5\text{-}74)$$

where $\qquad\qquad f^2 = R_x^2 + R_y^2 + R_z^2$

Once the vector $\mathbf{v}'$ is known, the vector $\mathbf{o}'$ is found from Eq. (5-73). Since the robot is of a cartesian type, the elements of $\mathbf{o}'$ are directly used as commands q_1, q_2, and q_3 to the X, Y, and Z actuators.

In order to derive the three wrist joint commands (q_4, q_5, and q_6) the three rotations ($\mathbf{R}_1$, $\mathbf{R}_2$, and $\mathbf{R}_3$) must be computed.

The three desired commands q_4, q_5, and q_6 can be related to the three rotation vectors by

$$f_i = \tan \frac{q_{i+3}}{2} \qquad (i = 1, 2, 3) \qquad (5\text{-}75)$$

and

$$\mathbf{R}_1 = \begin{bmatrix} f_1 \\ 0 \\ 0 \end{bmatrix} \qquad \mathbf{R}_2 = \begin{bmatrix} 0 \\ f_2 \\ 0 \end{bmatrix} \qquad \mathbf{R}_3 = \begin{bmatrix} 0 \\ 0 \\ f_3 \end{bmatrix} \qquad (5\text{-}76)$$

The three unknown rotation vectors describe the relative orientation between the tool and the base frame and the composite rotation must be equal to the desired tool orientation.

The composite rotation of $\mathbf{R}_1$ and $\mathbf{R}_2$ is $\mathbf{R}'$ and is obtained by Eq. (4-110) as

$$\mathbf{R}' = \mathbf{R}_1 \circ \mathbf{R}_2$$

$$= \frac{\mathbf{R}_1 + \mathbf{R}_2 + \mathbf{R}_1 \times \mathbf{R}_2}{1 - \mathbf{R}_1 \cdot \mathbf{R}_2} \qquad (5\text{-}77)$$

$$= \frac{\begin{bmatrix} f_1 \\ 0 \\ 0 \end{bmatrix} + \begin{bmatrix} 0 \\ f_2 \\ 0 \end{bmatrix} + \begin{bmatrix} f_1 \\ 0 \\ 0 \end{bmatrix} \times \begin{bmatrix} 0 \\ f_2 \\ 0 \end{bmatrix}}{1 - \begin{bmatrix} f_1 \\ 0 \\ 0 \end{bmatrix} \cdot \begin{bmatrix} 0 \\ f_2 \\ 0 \end{bmatrix}}$$

$$= \begin{bmatrix} f_1 \\ f_2 \\ f_1 f_2 \end{bmatrix}$$

the total composite rotation is

$$\mathbf{R}_1 \times \mathbf{R}_2 \times \mathbf{R}_3 = \mathbf{R'} \circ \mathbf{R}_3$$

$$= \frac{\begin{bmatrix} f_1 \\ f_2 \\ f_1 f_2 \end{bmatrix} + \begin{bmatrix} 0 \\ 0 \\ f_3 \end{bmatrix} + \begin{bmatrix} f_1 \\ f_2 \\ f_1 f_2 \end{bmatrix} \times \begin{bmatrix} 0 \\ 0 \\ f_3 \end{bmatrix}}{1 - \begin{bmatrix} f_1 \\ f_2 \\ f_1 f_2 \end{bmatrix} \cdot \begin{bmatrix} 0 \\ 0 \\ f_3 \end{bmatrix}} \tag{5-78}$$

$$= \frac{1}{1 - f_1 f_2 f_3} \begin{bmatrix} f_1 + f_2 f_3 \\ f_2 - f_1 f_3 \\ f_3 + f_1 f_2 \end{bmatrix}$$

Equation (5-78) describes the orientation of the tool after the three successive rotations of the wrist, which has to be equal to the desired orientation **R**

$$\mathbf{R} = \mathbf{R}_1 \circ \mathbf{R}_2 \circ \mathbf{R}_3 \tag{5-79}$$

or explicitly

$$\begin{bmatrix} R_x \\ R_y \\ R_z \end{bmatrix} = \frac{1}{1 - f_1 f_2 f_3} \begin{bmatrix} f_1 + f_2 f_3 \\ f_2 - f_1 f_3 \\ f_3 + f_1 f_2 \end{bmatrix} \tag{5-80}$$

Equation (5-80) presents three algebraic equations with three unknown variables f_1, f_2, and f_3. These variables are solved and used to obtain the commands q_4, q_5, and q_6.

$$q_4 = \tan^{-1} \frac{2(R_x - R_y R_z)}{1 + 2R_z^2 - f^2} \tag{5-81a}$$

$$q_5 = \sin^{-1} \frac{2(R_x R_z + R_y)}{1 + f^2} \tag{5-81b}$$

$$q_6 = \tan^{-1} \frac{2(R_z - R_x R_y)}{1 + 2R_x^2 - f^2} \tag{5-81c}$$

The definition of the initial orientation of the tool is established by the robot manufacturer. For the robot in Fig. 5-17 there are two logical initial conditions of the wrist, one shown in Fig. 5-17 and given in Eq. (5-76), and the other shown in Fig. 5-18. Its corresponding rotation vectors are

$$
\mathbf{R}_1 = \begin{bmatrix} f_1 \\ 0 \\ 0 \end{bmatrix} \qquad \mathbf{R}_2 = \begin{bmatrix} 0 \\ f_2 \\ 0 \end{bmatrix} \qquad \mathbf{R}_3 = \begin{bmatrix} f_3 \\ 0 \\ 0 \end{bmatrix} \qquad (5\text{-}82)
$$

The inverse kinematics solution for the first initial orientation [i.e., Eq. (5-76)] was already presented. The solution for the other initial orientation shown in Fig. 5-18 is derived in a similar way. The composite rotation is

$$
\mathbf{R} = \mathbf{R}_1 \circ \mathbf{R}_2 \circ \mathbf{R}_3 = \frac{1}{1 - f_1 f_3} \begin{bmatrix} f_1 + f_3 \\ f_2(1 + f_1 f_3) \\ f_2(f_1 - f_3) \end{bmatrix} \qquad (5\text{-}83)
$$

By comparing the $\mathbf{R}$ in Eq. (5-72) to Eq. (5-83) and substituting Eq. (5-75), the wrist variables are obtained

$$
q_4 = \tan^{-1} R_x + \tan^{-1}\left(\frac{R_z}{R_y}\right) \qquad (5\text{-}84a)
$$

$$
q_5 = 2 \tan^{-1} \sqrt{\frac{R_y^2 + R_z^2}{1 + R_x^2}} \qquad (5\text{-}84b)
$$

$$
q_6 = \tan^{-1} R_x - \tan^{-1}\left(\frac{R_z}{R_y}\right) \qquad (5\text{-}84c)
$$

As an example of the inverse kinematics solution, consider a cartesian robot with a wrist as illustrated in Fig. 5-18. Let us assume that the desired location of the TCP and the orientation of the tool are given by

$$
\mathbf{p} = \begin{bmatrix} 2 \\ 3 \\ 4 \end{bmatrix} \qquad \mathbf{R} = \begin{bmatrix} 0 \\ 1 \\ 0 \end{bmatrix}
$$

The tool vector $\mathbf{v}$ is initially $\mathbf{v} = [1, 0, 0]^T$. The problem is to derive the robot joint variables in order to achieve the desired position and orientation.

The solution consists of three parts: the computation of the desired position of point O', the derivation of the three prismatic joint variable values, and the computation of the three rotation joint values.

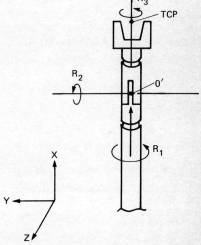

FIG. 5-18 Rotation vectors attached to the wrist's joints.

The required orientation of the tool is given by **R**, and the required tool vector **v**′ at its new position is calculated by Eq. (5-74)

$$\mathbf{v}' = \mathbf{v} + \frac{2\mathbf{R} \times (\mathbf{v} + \mathbf{R} \times \mathbf{v})}{1 + f^2}$$

$$= \begin{bmatrix} 1 \\ 0 \\ 0 \end{bmatrix} + \frac{2\begin{bmatrix} 0 \\ 1 \\ 0 \end{bmatrix} \times \left(\begin{bmatrix} 1 \\ 0 \\ 0 \end{bmatrix} + \begin{bmatrix} 0 \\ 1 \\ 0 \end{bmatrix} \times \begin{bmatrix} 1 \\ 0 \\ 0 \end{bmatrix} \right)}{1 + 1}$$

$$= \begin{bmatrix} 0 \\ 0 \\ -1 \end{bmatrix}$$

According to Eq. (5-73)

$$\mathbf{o}' = \begin{bmatrix} 2 \\ 3 \\ 4 \end{bmatrix} - \begin{bmatrix} 0 \\ 0 \\ -1 \end{bmatrix}$$

$$= \begin{bmatrix} 2 \\ 3 \\ 5 \end{bmatrix}$$

Because of the cartesian structure, the elements of **o**′ are used as the first three joint commands, namely

$$q_1 = 2 \qquad q_2 = 3 \qquad q_3 = 5$$

Utilizing Eqs. (5-84) the other three variables q_4, q_5, and q_6 are calculated

$$q_4 = 0 + 0 = 0$$

$$q_5 = 2 \tan^{-1} 1 = 90°$$

$$q_6 = 0 + 0 = 0$$

The results indicate that in order to achieve the required position and orientation, the six actuator commands are

$$\mathbf{q} = \begin{bmatrix} 2 \\ 3 \\ 5 \\ 0 \\ 90° \\ 0 \end{bmatrix}$$

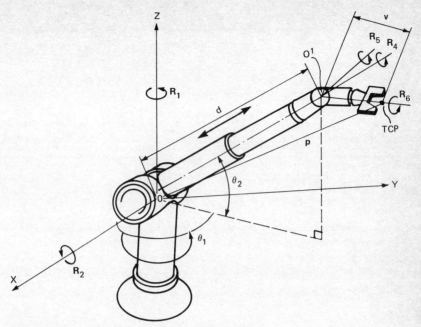

FIG. 5-19   A six-degrees-of-freedom spherical robot.

5.4.2 Inverse Kinematics for a Spherical Robot

A six-degrees-of-freedom spherical robot and the rotation vectors attached to its five rotating axes are shown in Fig. 5-19. The procedure to find the inverse kinematics solution of this type of robot is similar to that of the cartesian one. The computation of vector $\mathbf{o}'$ is exactly the same and it enables the calculation of the new position of the first three axes of motion as explained below. Using Eq. (4-3) the relation between the first three degrees of freedom and the vector $\mathbf{o}'$ is established

$$\mathbf{o}' = \begin{bmatrix} O'_x \\ O'_y \\ O'_z \end{bmatrix} = \begin{bmatrix} d_3 \cos \theta_1 \cos \theta_2 \\ d_3 \sin \theta_1 \cos \theta_2 \\ d_3 \sin \theta_2 \end{bmatrix} \qquad (5\text{-}85)$$

Equation (5-85) is solved for θ_1, θ_2, and d_3

$$\theta_1 = \tan^{-1} \frac{O'_y}{O'_x} \qquad (5\text{-}86a)$$

$$d_3 = \sqrt{O'^2_x + O'^2_y + O'^2_z} \qquad (5086b)$$

$$\theta_2 = \sin^{-1} \frac{O'_z}{d_3} \qquad (5\text{-}86c)$$

To find the other three joint variables, the required orientation is compared with the composite rotation

$$\mathbf{R} = \mathbf{R}_1 \circ \mathbf{R}_2 \circ \mathbf{R}_4 \circ \mathbf{R}_5 \circ \mathbf{R}_6 \qquad (5\text{-}87)$$

Equation (5-87) can be rewritten as

$$-\mathbf{R}_2 \circ -\mathbf{R}_1 \circ \mathbf{R} = \mathbf{R}_4 \circ \mathbf{R}_5 \circ \mathbf{R}_6 \qquad (5\text{-}88)$$

Since $\mathbf{R}_1$ and $\mathbf{R}_2$ are defined by the variables θ_1 and θ_2 that were previously determined, the left-hand side of Eq. (5-88) is known. According to Fig. 5-19, the five rotation vectors are

$$\mathbf{R}_1 = \begin{bmatrix} 0 \\ 0 \\ f_1 \end{bmatrix} \qquad \mathbf{R}_2 = \begin{bmatrix} 0 \\ -f_2 \\ 0 \end{bmatrix} \qquad \mathbf{R}_4 = \begin{bmatrix} f_4 \\ 0 \\ 0 \end{bmatrix}$$

$$\mathbf{R}_5 = \begin{bmatrix} 0 \\ f_5 \\ 0 \end{bmatrix} \qquad \mathbf{R}_6 = \begin{bmatrix} f_6 \\ 0 \\ 0 \end{bmatrix}$$

where

$$f_i = \tan \frac{\theta_i}{2} \qquad (i = 1, 2, 4, 5, 6)$$

The right-hand side of Eq. (5-88) was already calculated in Eq. (5-83), and the left-hand side is determined according to Eq. (4-110) in two steps

$$-\mathbf{R}_2 \circ -\mathbf{R}_1 = \cfrac{\begin{bmatrix} 0 \\ f_2 \\ 0 \end{bmatrix} + \begin{bmatrix} 0 \\ 0 \\ -f_1 \end{bmatrix} + \begin{bmatrix} 0 \\ f_2 \\ 0 \end{bmatrix} \times \begin{bmatrix} 0 \\ 0 \\ -f_1 \end{bmatrix}}{1 - \begin{bmatrix} 0 \\ f_2 \\ 0 \end{bmatrix} \cdot \begin{bmatrix} 0 \\ 0 \\ -f_1 \end{bmatrix}}$$

$$= \begin{bmatrix} -f_1 f_2 \\ f_2 \\ -f_1 \end{bmatrix}$$

and

$$-R_2 \circ -R_1 \circ R = \frac{\begin{bmatrix} -f_1 f_2 \\ f_2 \\ -f_1 \end{bmatrix} + \begin{bmatrix} R_x \\ R_y \\ R_z \end{bmatrix} + \begin{bmatrix} -f_1 f_2 \\ f_2 \\ -f_1 \end{bmatrix} \times \begin{bmatrix} R_x \\ R_y \\ R_z \end{bmatrix}}{1 - \begin{bmatrix} -f_1 f_2 \\ f_2 \\ -f_1 \end{bmatrix} \cdot \begin{bmatrix} R_x \\ R_y \\ R_z \end{bmatrix}}$$

$$= \frac{1}{1 + f_1 f_2 R_x - f_2 R_y + f_1 R_z} \begin{bmatrix} -f_1 f_2 + R_x + f_2 R_z + f_1 R_y \\ f_2 + R_y - f_1 R_x + f_1 f_2 R_z \\ -f_1 + R_z - f_1 f_2 R_y - f_2 R_x \end{bmatrix} \qquad (5\text{-}89)$$

The terms in Eq.(5-89) are known. Denoting the last composite vector by R''

$$R'' = -R_2 \circ -R_1 \circ R = \begin{bmatrix} R''_x \\ R''_y \\ R''_z \end{bmatrix}$$

Equations (5-89) are solved for the three joint variables q_4, q_5, and q_6

$$q_4 = \tan^{-1} R''_x + \tan^{-1} \frac{R''_z}{R''_y} \qquad (5\text{-}90a$$

$$q_5 = 2 \tan^{-1} \frac{\sqrt{R''^2_y + R''^2_z}}{1 + R''^2_x} \qquad (5\text{-}90b)$$

$$q_6 = \tan^{-1} R''_x - \tan^{-1} \left(\frac{R''_z}{R''_y} \right) \qquad (5\text{-}90c)$$

As an example, assume that a spherical robot (see Fig. 5-19) is required to locate the TCP in point P and to orient the tool according to R, where p and R are given by

$$p = \begin{bmatrix} 0 \\ 2 \\ 1 \end{bmatrix} \qquad R = \begin{bmatrix} 0 \\ 0 \\ 1 \end{bmatrix}$$

and the tool vector is $v = [1, 0, 0]^T$. Computing the vector v' according to Eq. (5-74) yields

$$\mathbf{v'} = \begin{bmatrix} 1 \\ 0 \\ 0 \end{bmatrix} + \frac{2\begin{bmatrix} 0 \\ 0 \\ 1 \end{bmatrix} \times \left(\begin{bmatrix} 1 \\ 0 \\ 0 \end{bmatrix} + \begin{bmatrix} 0 \\ 0 \\ 1 \end{bmatrix} \times \begin{bmatrix} 1 \\ 0 \\ 0 \end{bmatrix} \right)}{1 + 1} = \begin{bmatrix} 0 \\ 1 \\ 0 \end{bmatrix}$$

Subsequently the vector $\mathbf{o'}$ is calculated according to Eq. (5-73)

$$\mathbf{o'} = \mathbf{p} - \mathbf{v'} = \begin{bmatrix} 0 \\ 2 \\ 1 \end{bmatrix} - \begin{bmatrix} 0 \\ 1 \\ 0 \end{bmatrix} = \begin{bmatrix} 0 \\ 1 \\ 1 \end{bmatrix}$$

The first three joint variables are determined from Eq. (5-86)

$$q_1 = \theta_1 = \tan^{-1} \frac{1}{0} = 90°$$

$$q_2 = \theta_2 = \sin^{-1} \frac{1}{\sqrt{2}} = 45°$$

$$q_3 = d_3 = \sqrt{0^2 + 1^2 + 1^2} = \sqrt{2}$$

The rotations $\mathbf{R}_1$ and $\mathbf{R}_2$ are derived from q_1 and q_2 as follows:

$$f_1 = \tan \frac{q_1}{2} = 1 \qquad f_2 = \tan \frac{q_2}{2} = \frac{1}{1 + \sqrt{2}}$$

$$\mathbf{R}_1 = \begin{bmatrix} 0 \\ 0 \\ f_1 \end{bmatrix} = \begin{bmatrix} 0 \\ 0 \\ 1 \end{bmatrix} \qquad \mathbf{R}_2 = \begin{bmatrix} 0 \\ -f_2 \\ 0 \end{bmatrix} = \begin{bmatrix} 0 \\ -\dfrac{1}{1 + \sqrt{2}} \\ 0 \end{bmatrix}$$

The required orientation $\mathbf{R}$ and the calculated rotations $\mathbf{R}_1$ and $\mathbf{R}_2$ enable to calculate the wrist joint variables from Eq. (5-89).

$$\mathbf{R}'' = \frac{1}{1 + 1} \begin{bmatrix} +\dfrac{1}{1 + \sqrt{2}} - \dfrac{1}{1 + \sqrt{2}} \\ -\dfrac{1}{1 + \sqrt{2}} - \dfrac{1}{1 + \sqrt{2}} \\ -1 + 1 \end{bmatrix} = \begin{bmatrix} 0 \\ -\dfrac{1}{1 + \sqrt{2}} \\ 0 \end{bmatrix}$$

the joint variables q_4, q_5, q_6, can now be calculated using Eq. (5-90)

$$q_4 = \tan^{-1} 0 + \tan^{-1} 0 = 0°$$

$$q_5 = 2 \tan^{-1} \frac{\sqrt{[1/(1 + \sqrt{2})]^2 + 0^2}}{1 + 0^2} = 45°$$

$$q_6 = \tan^{-1} 0 - \tan^{-1} 0 = 0°$$

The robot six joint variables are

$$q = \begin{bmatrix} 90° \\ 45° \\ \sqrt{2} \\ 0° \\ 45° \\ 0° \end{bmatrix}$$

Note that according to the demands the TCP is located in the YZ plane and the tool vector is parallel to the Y axis.

REFERENCES

Amitai, G.: "Structural Analysis and Trajectory Computation Viewing Adaptive Robots," Ph.D. Thesis, Technion, Israel Institute of Technology, Dept. of Mechanical Engineering, 1984.

Anderson, T. R.: "High Speed Coordinate Control of Industrial Robots," *Proc. 9th ISIR*, 1979, pp. 441–461.

Bong, J. L., and S. N. Tae: "Interpolation for 2-Dimensional Contouring Control," *Proc. 13th ISIR*, April 1983, Sec. 13, pp. 121–126.

Brady, M. et al.: *Robot Motion, Planning and Control*, MIT Press, Cambridge, Mass., 1982.

Drazan, P. J. et al.: "The Use of Interpolation Routines for Path Generation and Control of Electro-Pneumatic Industrial Manipulator," *Proc. 10th ISIR*, 1980, pp. 313–330.

Featherstone, R.: "Position and Velocity Transformations between Robot End Effector Coordinates and Joint Angles," Int. J. Robot. Res. 2, 1983.

Holm, R. E.: "Computer Path Control of an IR," *Proc. 8th ISIR*, 1978, pp. 327–332.

Konstantinov, M. S., and S. P. Patarinski: "A Contribution to the Inverse Kinematic Problem for Industrial Manipulators," *Proc. 12th ISIR*, 1982, pp. 459–468.

Lee, C. S. G., and M. Ziegler: "A Geometric Approach in Solving the Inverse Kinematics of Puma Robots," *Proc. 13th ISIR*, April 1983, Sec. 16, pp. 1–15.

Lewis, R. A.: "Autonomous Manipulation on a Robot, Summary of Manipulator

Software Functions," Jet Propulsion Laboratory, California Institute of Technology, TM 33-679, March 1974.

Lien, T. K.: "Coordinate Transformation in CNC Systems for Automatic Handling Machines," *CIRP Manuf. Syst.*, vol. 9, no. 1, 1980, pp. 49–60.

Limken, G. A. et al.: "Control of Working Organ Movement Speed in the Manipulator Nonorthogonal System of Coordinates," *Proc. 8th ISIR*, 1978, p. 600.

Mancini, L. et al.: "A New Approach to Coordinate Transformation: Parallel Computation," *Proc. 13th ISIR*, April 1983, Sec. 16, pp. 19–30.

Paul, R. P.: *Robot Manipulators, Mathematics, Programming and Control*, MIT Press, Cambridge, Mass., 1981.

Paul, R.: "Modelling, Trajectory Calculation and Servoing of a Computer Controlled Arm," Stanford University, AI Laboratory, Memo AIM 177, 1972.

Paul, R.: "Manipulator Cartesian Path Control," *IEEE Trans. Syst. Man Cybernet.*, SMC-9 1979, pp. 702–711.

Pieper, D. L.: "The Kinematics of Manipulators under Computer Control," Ph.D. Thesis, Stanford University, Dept. of Mechanical Engineering, October 1968.

Renaud, M.: "Robot Manipulator Control," *Proc. 9th ISIR*, 1979, pp. 463–475.

Shepperd, S. W.: "Quaternion from Rotation Matrix," *J. Guid. Control*, vol. 1, no. 3, 1978, pp. 223–224.

Shoham, M.: "Development of Algorithms for Robot Spatial Motions," M.Sc. Thesis, Technion, Israel Institute of Technology, April 1982.

Taylor, R. H.: Planning and Execution of Straight Line Manipulator Trajectories, *IBM J. Res. Dev.*, vol. 23. no. 4, 1979, pp. 424–436.

Waters, R. C.: "Mechanical Arm Control, AI Laboratory, Massachusetts Institute of Technology, AIM 549, October 1979.

Whitney, D. E.: "The Mathematics of Coordinate Control of Prosthetic Arms and Manipulators," *J. Dyn. Syst. Meas. Control*, December 1972, pp. 303–309.

Zaprjanov, J., and S. Boeva: "Hierarchical Decentralized Control of Industrial Robots," *Proc. 8th Triann. World Cong.*, Kyoto, Japan, August 1981, pp. 129–134.

CHAPTER 6

Applications of Robots

Industrial robots are primarily installed today to improve productivity and product quality. The main applications of industrial robots are

Loading and unloading machine tools
Handling in manufacturing processes, such as die casting
Welding
Spray painting
Assembly
Machining, such as deburring and drilling
Inspection

These applications are discussed in the present chapter.

6.1 HANDLING, LOADING, AND UNLOADING

During the sixties and the seventies automation has primarily affected the manufacturing process and machine tool control but not the auxiliary functions such as handling, setup, loading, and unloading. The time spent to transfer a workpiece from one station to the next is still high. Up to 95 percent of the time involved in manufacturing a part is composed of transfer and waiting time, and only about 5 percent of the total

FIG. 6-1   A robot designed for loading and unloading machine tools. (FANUC.)

time is the part in an actual processing stage (Bjorke, 1979). Whereas the processing time has been reduced considerably by automation, much less progress has been made in handling and loading. The fully automatic systems that were developed for mass production (e.g., transfer lines in the automobile industry) are rigid (hard automation) and not suitable for batch production (series on the order of 50 to 100,000 parts annually). A more flexible automation technology which takes into account frequent changes in production is needed for this category of industrial production, which accounts for about 75 percent of the manufactured parts.

A new solution is offered to the handling and machine tool loading of small and medium series of parts with the development of industrial robots. Actually, loading and unloading machine tools is one of the major applications of robots, as seen in Fig. 1-2. Robots are utilized to load and unload machine tools in two basic configurations: (1) a robot tending a single machine, (2) a robot serving several machines. The first configuration is applied when typical machining times per part are short, and the second system, when a chain of operations must be executed to complete a part.

6.1.1   Single Machine Tool

The automation of a machine shop might start by applying a robot to load and unload one machine tool in the shop. Because of the structure of machine tools, robots with cylindrical and spherical coordinate systems are best suited for this application. Such a robot is shown in Fig. 6-1. Since the loading is assisted by mechanical fixtures, the required repeatability is relatively low, and a range of 0.2 to 0.5 mm is usually satisfactory.

The machine tool which is selected for automation must be of an NC or CNC type (see Sec. 1.5), since the robot computer must communicate with the machine controller. The robot computerized controller sends a signal to the machine when the loading is completed, and the machine controller transmits a signal to the robot when machining is finished, thereby synchronizing the operation of the two controllers. To facilitate the synchronization, it is often advantageous if both the CNC and the robot controller are produced by the same manufacturer. Such a configuration is shown in Fig. 6-2, where the robot in Fig. 2-22 has been interfaced with a small CNC machining center supplied by the robot's manufacturer.

The total production time of a part for any machining operation is comprised of three individual times: (1) Actual cutting time, t_1; (2) tool-changing time (the relative portion per part), t_2; (3) loading and unloading time, t_3.

The cutting time may be reduced by using better tools and applying adaptive control and optimization techniques (Hitomi, Nakajima, and

FIG. 6-2 A robot serves a CNC milling machine, both of the same manufacturer. (Shanoa Electronics.)

Osaka, 1978). The introduction of the NC and CNC has mainly affected the cutting process itself, but not the auxiliary functions such as tool changing, loading, and unloading. Further reduction in production time can be achieved if tool changing and loading and unloading are performed by robots. Tool changing, however, is regarded as too complex an operation and usually is not performed by robots. Loading raw materials onto the machines and unloading the finished parts are relatively simple operations. The robot picks up the workpiece, places it in the machine chuck, waits until the end of the cutting operation, and then unloads the finished part. The efficiency of the robot operation e can be defined as the ratio of the time saved to the total production time:

$$e = \frac{t_3 - t_4}{t_1 + t_2 + t_3}$$

where t_4 is the sum of the robot's loading and unloading times. The times t_2, t_3, and t_4 are essentially constants, so the efficiency mainly depends on the actual cutting time t_1. Usually t_2, t_3, and t_4 are each much less than 1 min, and if the cutting time exceeds 1 or 2 min it is not economical to dedicate a robot to loading and unloading of a single machine tool. In such a case other tasks can be assigned to occupy the robot while it waits for the cutting, or a single robot can be installed to serve the loading and unloading of several machine tools, a configuration which is known as a *manufacturing cell.*

6.1.2 Several Machines

A single robot can be installed to serve several machine tools. A spherical coordinate robot which loads and unloads two multispindle machine tools in the manufacture of cylinder heads is shown in Fig. 6-3. This robot handles bars that weigh up to 23 kg with a repeatability of 0.2 mm.

The production of most parts requires several machining operations which are performed by different machines. Several machines (typically between two and four) can be grouped into a cell and arranged around a robot which loads and unloads the machines. A workpiece is loaded by the robot for processing in the first machine tool, and thereafter is transferred from one machine to the next, until the processing of the part is completed. The finished part is subsequently unloaded by the robot from the last machine tool. An automatic inspection station may also be added to the cell to measure the dimensions of the finished part. The production process in the cell is controlled and monitored by a computer dedicated to these tasks.

FIG. 6-3   A robot loads and unloads two multispindle machine tools. (Prab.)

The increasing role of the manufacturing cell in the production justifies a special consideration of this subject.

6.2 THE MANUFACTURING CELL

The production of parts in several industries is performed today in manufacturing cells. Each cell consists of several machine tools served by a single robot. The robot is usually equipped with a double gripper and therefore can unload a finished part and load a new workpiece without leaving the machine tool bed. It operates as follows: The robot first picks up a raw workpiece with one gripper, then moves to the first machine tool, unloads the semifinished part from the machine with its other gripper, rotates the wrist 180°, and loads the new workpiece while still holding the semifinished part. Next, the robot moves to the second machine tool, unloads the part from the second machine with the free gripper, and then loads the semifinished part taken from the first machine for continuation of machining. This procedure continues on through every machine tool in the cell. At the end of the process, the finished part is unloaded by the robot from the last machine tool and placed on an outgoing conveyer. The robot then picks up a new raw workpiece to start another cycle of the process.

6.2.1 The Cellular Concept

The manufacturing cell concept was introduced by the Production Engineering Laboratory at the University of Trondheim in Norway (Bjorke, 1979; Koren, 1983; Merchant, 1980). To understand the cellular concept it is necessary to understand certain current economic and social conditions in Norway. Norway is a country that has a small and scattered population, which is desirable for national security reasons. The rural areas of Norway suffer from severe unemployment, while in its cities there is a growing labor shortage. For security and economic reasons, the Norwegian government does not want the rural population to move to the cities. The proposed solution was to bring miniplants of manufacturing to the rural areas. This was the original cellular concept.

Under this cellular concept, the manufacturing operations are broken down into separate cells, each at a different miniplant. Each cell is responsible for the manufacture of a specific part family, namely, parts with similar geometric features as determined by group technology principles. The cells are interconnected by a network of material and finished parts transports.

However, Norwegian wages are among the highest in the world. Therefore, in order for the Norwegian manufacturing industry to be competitive in the world market, its productivity must be high. This brought the second concept in the plan: to provide each miniplant with a high-technology core.

The core consists of a group of several (two to five) CNC machine tools arranged in a circle around a single robot, as shown in Fig. 6-4. The robot does all the part handling and machine loading and unloading in the cell. The supervision and the coordination among the various operations is performed by the cell computer. The core runs 24 hours continuously but requires workers' participation only during the day. The day shift prepares the computer programming, production planning, scheduling, heat treatment, mounting parts on pallets, etc.

The utilization of the cellular concept does not necessarily require that the manufacturing cells be located great distances from each other. On the contrary, a higher productivity can be achieved if all the cells are located along a single material transfer system, such as a long conveyor, on which raw workpieces and semifinished and finished parts are moving. A "ready for workpiece" signal from the control unit of the first machine in a manufacturing cell instructs the robot to look for the required workpiece on the conveyor. The robot picks up the workpiece, loads it onto the machine, and sends a signal to the machine control to begin its operations on the workpiece. While waiting for com-

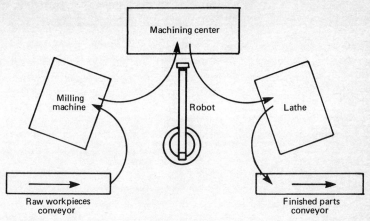

FIG. 6-4   A manufacturing cell consisting of a robot and three machine tools.

pletion or transfer of the part to the next machine, the robot performs tool-changing and housekeeping functions such as chip removal, staging of tools in the tool changer, and inspection of tools for breakage or excessive wear. For each of these chores, other interconnect signals can alter the functions performed by the robot depending on the outcome of these tests or the presence of any unusual situations encountered during the housekeeping. If during these functions a machine control detects a malfunction or a tool breakage during a machine operation, the robot must abandon these routine tasks and take some action to either remedy the problem or initiate an emergency procedure for the total system. A "part finished" signal from the last machine tool to the robot requests that the finished part be unloaded and transferred to the outgoing conveyor. The cycle is then repeated.

6.2.2 The State of the Art

Manufacturing cells, or machining cells, are in operation in industry. In FANUC Inc., a Japanese producer of CNC controllers and robots, the production has been performed in cells since the late seventies (Koren, 1983). A typical cell consists of two or three machine tools and a rotary indexing table, arranged in a circle around a robot. Each indexing table contains initially raw workpieces. The robot picks up a workpiece, loads it onto the first machine, and later transfers it between the machines for processing. At the end of the process the robot returns the finished part to the same location on the indexing table from which the workpiece was taken. The table then moves one station and the cycle is repeated. At the end, the indexing table contains only finished parts. Details about FANUC's Fuji factory are given in Chap. 10.

In the United States, Wasino Corp. installed a cell consisting of two CNC lathes and a double-gripper robot (Stauffer, 1981a). In a typical sequence, the robot picks up a workpiece from a table with one of its two grippers and approaches the spindle of the first machine. There the robot retrieves the partially machined workpiece from the spindle with its second gripper, rotates its wrist 180°, and substitutes the new workpiece from its other gripper. The robot then moves to the second machine and substitutes the finished part at the spindle with the partially machined workpiece which was in its gripper. Finished parts are lifted out by the robot and deposited in a hopper.

International Harvester's Farmall Division installed a similar manufacturing cell consisting of a heavy-duty industrial robot, two turning centers, and a gauging station. Both the robot and the machines are produced by the same manufacturer, which facilitates the interfacing among the various controllers in the cell. Parts enter the cell as rough castings and leave it completely machined and inspected. The robot and one turning center of the cell are shown in Fig. 6-5.

FIG. 6-5   A robot serves a turning center in a manufacturing cell. (Cincinnati Milacron.)

The sequence of operations in the cell is as follows:

1. The robot takes a part from the pallet on the conveyor to the first turning center (not shown). Entering from the rear of the machine, the robot removes the part just machined and loads the part from the pallet. Loading the turning centers from the rear leaves the front of the machines open for operator observation, control access, and tool replacement.

2. The robot takes the part it just removed from the first machine to an automatic gauging station. If the part is found to be within tolerances at this point, the robot takes it to the second turning center (as shown above).

3. The robot removes a finished part from the second turning center and loads the part just gauged for a second turning operation.

4. The robot takes the finished part to the gauging station. If the part is good, it is returned to the pallet, and the robot picks up another part and the sequence is repeated.

An advanced manufacturing cell for wheel hub manufacturing was installed by Olofsson at Deer and Co. (Stauffer, 1981b). The cell is shown in Fig. 6-6 and consists of the following:

1. One programmable robot with end-arm tooling capable of handling a family of hubs. The robot loads, transfers, and unloads two hubs at once between the stations.

2. Two Model 75 CNC vertical twin-spindle turning and boring machines. Both are robot-loaded.

3. One part orientation station which locates hubs for proper drilling and tapping on the head changer.

4. One Olofsson head center with a capacity for drilling, tapping, and boring several positions in one pass. This machine automatically changes four multispindle heads for drilling and tapping from a turret-type indexing unit. The head change is accomplished in 28 s. Four heads accommodate the entire hub family.

5. Two conveyor tracks for incoming and outgoing parts.

The system uses the robot to remove parts from the incoming conveyor, load the machines—each accepting parts in a different attitude—and to place the completed parts on the outgoing conveyor. Robot loading, transfer, and unloading make possible a job which no single operator could accomplish. In the operation where the hubs

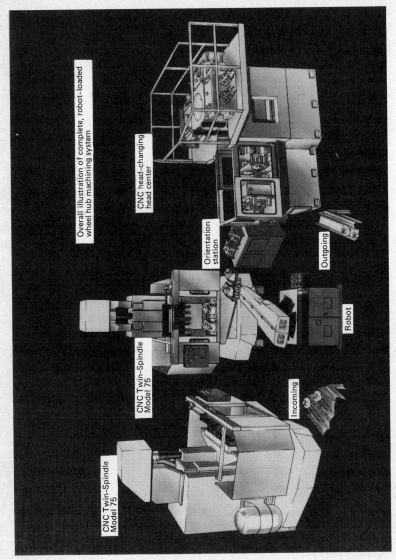

Overall illustration of complete, robot-loaded wheel hub machining system

CNC head-changing head center

Orientation station

Outgoing

CNC Twin-Spindle Model 75

Robot

CNC Twin-Spindle Model 75

Incoming

FIG. 6-6 Illustration of a wheel hub machining cell. (Olofsson.)

weigh 40 lb each, the robot handles 80 lb at each station, since the system is designed for the simultaneous machining of two hubs.

The robot will, in completing its movements between the system's six stations, transfer and handle over 11,000 lb in 1 hour.

The operation sequence for the system is as follows: the robot opens the guard door on the head center, unloads two parts, and places them in the outgoing conveyor. Next the robot opens the guard doors on the second Model 75 and unloads the two parts, rotates them 90°, and loads the head center. Guard doors are closed and the head center cycle begins. Guard doors are then opened on the first Model 75 and the two parts are unloaded, rotated 180°, and the second Model 75 is loaded. After closing the guard doors, the cycle begins for the second Model 75. Next the robot picks up two parts from the incoming conveyor, loads the first Model 75, and closes the guard doors for the cycle to begin. Finally, the robot returns to the head center to await the end of the cycle. Upon receiving the appropriate signal the procedure is repeated.

Total cycle time for the system is 128.6 s. Load-unload time is 30 s. Resultant production is 56 hubs per hour.

One of the most advanced manufacturing cells was installed at Xerox Corporation in the United States (Engelberger, 1980). It consists of a transfer conveyor along which three robots serve a variety of machines: two CNC lathes; brazing, grinding, and broaching machines; and an inspection station.

It should be noted that robots in manufacturing cells must accommodate a great variety of loading and unloading paths and therefore must have six degrees of freedom. The coordinate system might be articulated, spherical, or cylindrical, but not cartesian. The robot must have a long enough reach to serve all machines and be able to carry the heaviest parts being handled in the cell.

6.2.3 Optimization of the Production Rate

The optimization of the production rate in a manufacturing cell requires a different strategy than the one applied to a single machine tool. Methods to achieve maximum productivity of one machine tool are well known [e.g., (Wu and Ermer, 1966)]. However, applying them to each machine in the cell does not guarantee maximum productivity of the cell. If each machine in the cell is optimized independently, then either machines would be idle or parts would be waiting to be machined, and an overall cellular maximum production rate would not be achieved.

For a manufacturing cell, the correct strategy is to optimize the cutting speed of the slowest machine by using the conventional minimum production time criterion. Thereafter, the production rate of the other machines in the cell should be synchronized with the slowest one in order to obtain an equal cycle time on each machine (Hitomi, Nakajima, & Osaka, 1978). With this strategy the overall product cycle time is the sum of the machining time, plus a relative portion of the tool-changing time, plus workpiece loading and unloading times.

6.3 WELDING

Welding is a manufacturing process in which two pieces of metal are joined usually by heating them until molten and fused and/or by applying pressure. The welding operations performed by robots are thermal processes in which the metals are joined by melting or fusing their contacting surfaces. These processes can be grouped under two classes: (1) autogenous welding, in which no filler material is added to the joint interface, and (2) homogeneous welding, in which a filler material of the same type as the parent metal is added. Accordingly, there are two types of welding operations performed by robots: spot welding and arc welding. Spot welding is an autogenous process in which the two pieces of metal are joined only *at certain points.* The required heat is generated by the passage of an electric current through the metals at the point where they are to be joined. Spot welding is frequently used in the automotive industry to join thin sheet metals. Arc welding is a homogeneous process in which the two metals are joined *along a continuous path.* The required heat is provided by an electric arc generated between an electrode and the metals. Arc welding is needed, for example, in sealing a container against leakage. Each of these welding operations requires a different type of equipment and a different control system for the robot arm, as is further discussed below.

6.3.1 Spot Welding

In spot welding, two pieces of metal are touching one another at the required joint point and then a large alternating current (for example, 1000 A) is transferred through the metals during a short period t. The welding current is transferred through two nonconsumable electrodes which are pressed against the two metals. Since an electric resistance R exists at the touch point, the current I generates a heat q given by

$$q(t) = \int_0^t I^2 R(t) \, dt \qquad (6\text{-}1)$$

Note that R is strongly dependent on the welding temperature and is increased, for example, by a factor of 10 as the temperature rises from room temperature to 1000°C (1800°F). Part of the generated heat melts the two pieces of metal and fuses them together. The welding current is then turned off, but the electrodes are kept closed in order to let the weld cool. This process is repeated at each point to be welded. The thickness of the metal pieces should be small enough to avoid conducting heat away from the welded points, and therefore spot welding is usually used only for joining metal sheets.

From the above discussion it is seen that four major parameters affect the quality of the welded spots:

1. The magnitude of electric current
2. The pressure provided by the electrodes (the resistance between the two metals is inversely proportional to this pressure)
3. The time during which the current is supplied
4. The thickness and material of the welded metals

The first three parameters must be correctly selected to suit the thickness and type of the welded material. In a welding robot system these parameters are set initially and maintained throughout the work. As a consequence, repeated production conditions are obtained which result in high-quality welds.

Features of Spot-Welding Robots.   The spot-welding robot has to carry the welding gun, which consists of the electrodes, the cables which are required to conduct the high current, and sometimes a water-cooling system for the electrodes. The welding gun is relatively heavy (10 to 80 kg), and many dc-motor-driven robots cannot handle such heavy loads. Therefore, most of the spot-welding robots are hydraulically powered.

The control system for spot-welding robots is of a PTP type. The desired positional accuracy is usually not high, and a positional repeatability of ±1 mm is sufficient. This repeatability is much better than that obtained by human welders. Further, the operation of robotized spot welding is very fast. When the distance between spot welds is an inch or two, several spot welds can be made per second—faster than human welders. Positioning of the welds is more accurate, resulting in more uniform quality.

Spot welding generates sparks which might be detected by the robot controller as feedback pulses representing BRUs. Therefore, robots operating in a spot-welding environment may require isolation transformers or special screening and filtering devices for their controllers to protect them from the electrical noise and ensure reliable positioning. Another special feature is an arm that encloses and thereby protects the cables from sparks.

Spot-Welding Applications.   The 12th American Machinist Inventory of Metalworking Equipment (1976–1978) reveals that 58,000 spot welders are in use in the United States (Jablonowski, 1983). They are used in fabrication of structural metal products, domestic appliances, metal furniture, containers which do not require liquid-tight joints, etc. Spot-welding robots are gradually penetrating into these industries. However, nowhere have spot-welding robots affected industry operation more than in car body assembly.

The first spot-welding robots were installed in 1969 at a General Motors plant for welding car bodies (Engelberger, 1980). Since then spot-welding robots have proved to be very profitable and a few thousand are in use today in the automotive industry. Figure 6-7 shows a typical operation of spot-welding robots on an automobile assembly line. Parts of the robots are suspended from the ceiling, which saves expensive floor space. Several robots can operate simultaneously on the

FIG. 6-7   Robots perform spot welding in an automobile assembly line. (Unimation.)

FIG. 6-8 Two industrial robots tack-welding parts of the Ford Escort subassembly (Kuka.)

same car body, which increases the efficiency of the assembly line. Better efficiency is also obtained by specifying fewer welds for robot welding than with human welders. A human operator might miss a weld or make it in an incorrect location, and therefore many times extra welds are added at the design stage. With robot operation the work is consistent and all the welds are placed in the right location and therefore the required body strength can be achieved by specifying fewer welds.

Much of the work currently done by robots in the automotive industry is in the area of respotting (Engelberger, 1980). This is a technique where an automobile body is simultaneously welded by several robots on the production line. These robots are under the control of a supervising computer which signals the arrival of a particular body style and causes the robots to switch to the appropriate task program in their memory systems. Figure 6-8 shows two robots tack-welding parts on the Ford Escort line in Europe. They can place spot welds into a random mix of two- or four-door body styles. Sensing devices (such as vision) can easily be incorporated to check the body style that the robots are expected to weld in order to avoid catastrophic mistakes.

The spot-welding robots in the automotive industry are not simple

PTP systems. A typical assembly line produces between 50 and 90 cars per hour, and the work is performed while the car bodies are continuously moving on conveyors, which means that the weld locations specified by the task programs should be synchronized with the velocity of the assembly line. Since the velocity of the assembly line is constant, the positional compensation is inserted off line into the task programs, and consequently those programs are appropriate only for one specific line velocity.

6.3.2 Arc Welding

Arc welding is a method used for joining metals along a continuous path by the heat derived from an electric arc. Arc welding includes such methods as conventional welding with carbon and metal electrodes, submerged arc welding, inert gas welding, and atomic hydrogen welding.

Arc welding, as commonly practiced, requires the use of a welding generator to supply the electric power to produce the arc. The generator is usually of a dc type, although an ac generator is sometimes used. As is seen in Fig. 6-9, one lead of the generator is connected to the electrode holder that holds either a permanent (e.g., carbon) or a consumable (metallic) rod, and the second lead is attached to the work to be welded. For welding with uncoated electrodes with a dc generator, the lead to the electrode is usually from the negative pole of the generator, and the lead to the work is from the positive pole as shown in Fig. 6-9. In other cases, where coated metal electrodes or inert gas welding methods (see below) are used, the polarity may be reversed.

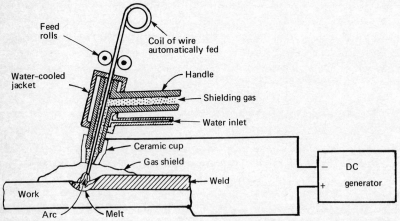

FIG. 6-9   Inert gas shielded arc welding with a continuous wire fed as electrode (J. Datsko.)

By holding the electrode an appropriate distance from the work, the operator is able to produce an arc discharge between the electrode and the work. Welding heat is provided by the electric arc, and temperatures of approximately 6500°C (~12,000°F) are achieved.

The original arc-welding method was carbon arc welding, which employed a carbon rod as the negative electrode in making the arc at the weld. Today, tungsten rather than carbon is the most common electrode material in this process. The heat of the arc melts a pool of metal on the surface of the work. If additional metal is required, it is supplied by melting a separate metallic rod into the weld.

Metal arc welding is similar to carbon arc welding except that the permanent electrode is replaced by a consumable metal electrode in the form of a rod or wire that is fed and melted into the weld, thereby acting as its own filler metal. The composition of the metal electrode is usually equal to or of a richer alloy mixture than the metal being welded.

The nonconsumable electrode process is not used nearly as extensively as the consumable electrode process. Bare rods, however, are seldom used as electrodes. A bare wire in the form of a coil having the same composition as the material to be welded is used for much of the production welding. The disadvantage of the bare electrode process is that the weld pool is unprotected from the oxygen in the atmosphere and so the welds are very inferior due to the presence of oxide inclusions in the weld zone.

Shielded arc and inert gas welding are arc-welding procedures used when oxidation of metals must be considered. The natural affinity of most metals for the oxygen in the air causes them to oxidize with increased rapidity as temperatures increase. If such oxidation of metals being welded is not critical to the welded structure, it is not necessary to guard against oxidation, and the electrodes, either carbon or metal, may be used bare. If, however, oxidation is detrimental to the welded structure, an inert gas must be introduced that will suppress the formation of oxides. This is accomplished by four major methods:

1. Coating the electrodes with a flux material that, when heated, will give off a gas such as CO_2 in an amount sufficient to prevent the association of the metal with oxygen. When the rod is used up another is inserted. Coated metal electrodes are used almost exclusively for job shop work, such as repairs, and in portable welding units. With coated electrodes it is possible to obtain weldments free of oxide inclusions.

2. The inert gas shielded (IGS) method, in which the arc shielding is accomplished by introducing a flow of inert gas (helium, argon, or

carbon dioxide) at the weld and using either a permanent electrode (such as one made of tungsten) with a filler metal in the weld or a continuous metal wire coiled on a drum. The gas continuously protects the weld from the atmosphere to inhibit oxidation. Inert gas welding is sometimes known under the trade names metal inert gas (MIG) (when metallic electrode is used) and tungsten inert gas (TIG) (when tungsten electrode is used).

3. Submerged arc welding is another form of shielded welding, wherein oxidation is prevented by covering the weld area with a flux. The flux is continuously fed through a pipe directly in front of the arc, so that the arc and all the adjacent hot metal are covered and protected from the oxygen in the air. The filler metal is a long bare wire in the form of a coil.

4. Atomic hydrogen welding, which maintains an arc between two tungsten electrodes, around which flows a stream of hydrogen gas. The heat of the arc transforms the molecular hydrogen to atomic hydrogen with resulting higher temperatures than those obtained by the metal arc method, plus perfect shielding at the weld.

The IGS arc with a consumable wire electrode is the process applied in most arc-welding robotic systems (Jablonowski, 1983). The advantage of this process over the coated electrodes is mainly that there is no slag to clean off after welding. The higher welding speeds also partially offset the higher cost of the process. The process originally used helium or argon gas and therefore was very expensive. However, with the introduction of carbon dioxide as a shielding gas, the IGS process replaced the coated electrode process in many steel welding operations, particularly in the automotive and appliance industries (Datsko, 1979), and subsequently this process became the most commonly used one with arc-welding robots.

While most robotic arc welding uses a consumable wire electrode (i.e., MIG welding) with an automatic wire feeder, welding with non-consumable tungsten electrodes with shielding gas (i.e., TIG welding) is also in use. Two robots equipped for IGS arc welding are shown in Fig. 6-10. The robots are set up to perform separate operations on a stainless steel part. One unit is set up for TIG welding, while the other is set up for MIG welding. The control unit of each of these welding robots can be interfaced also with a positioning robot so that the port can be continuously manipulated during the welding process.

In arc welding the robot uses the welding gun as a tool. The mechanism is demonstrated in Fig. 6-9. The consumable electrode, which provides the filler material, is in the form of a wire (coiled on a drum)

FIG. 6-10   Two six-axis robots do arc welding. (Advanced Robotics.)

of the same composition as the material to be welded. Wire diameters
of $\frac{1}{32}$ to $\frac{3}{16}$ in (0.8 to 4.8 mm) are commonly used. The wire is automatically fed by a motor with adjustable speed at a preset rate that is determined by the arc voltage. The wire feed increases with an increase in
the voltage applied between the work and the electrode. This voltage
can be monitored and used to maintain a constant arc length by varying
the speed of the motor which feeds the wire. In order to keep the electrode cooler and permit higher currents to be used, the shielding gas
flows in a tube along the electrode. The tube is terminated in a nozzle
at the end of the gun from which the gas flows into the arc region.
Welding robotic systems sometimes use water-cooled guns.

The weight of the welding gun is usually not heavy (unless the water-
cooled type is used) and therefore dc servomotor-driven robots are typically used in arc welding (see Fig. 6-11), although hydraulically driven
robots are also sometimes found. Welding speeds range from about 10
to over 120 in/min (0.25 to 3 m/min). The welding current usually
ranges between 100 and 300 A, but with the larger electrodes ($\frac{3}{16}$ in)
the current may be as high as 1200 A (Datsko, 1979), resulting in a very
deep penetration of the weld. The control of both the rate at which the
wire electrode is fed and the welding cycle (i.e., the time during which

FIG. 6-11   A dc servomotor-driven, six-axis robot designed for process applications such as arc welding (Cincinnati Milacron.)

the inert gas flows) are performed by the standard welding equipment. The task of the robot is to lead the welding gun along the programmed trajectory.

The control system for robots in arc welding is usually of a CP type. Nevertheless, PTP control systems are also used. In PTP programming the required trajectory is divided into a large number of small (e.g., 1 cm) and equal segments. In all cases the control computer of the robot is interfaced with the control unit of the welding equipment in order to synchronize the start and termination of the robot motions with the cycle of the welding equipment.

6.3.3   Comparison

Both spot welding and arc welding are manufacturing processes by which metals are joined. In both processes the robot guides the welding gun to or along the work. However, as shown in Table 6-1, the two welding robotic systems are not as similar as they might appear. Each requires different welding equipment and a different robotic control system. The spot-welding robot is designed to carry a heavy welding gun from point to point and is usually hydraulically powered. By contrast, the arc-welding robot carries a much lighter gun at constant speed along a continuous path and usually employs electrical motors as the drive elements.

TABLE 6-1 Comparison Between Spot-Welding and Arc-Welding Robots

	Spot welding	Arc welding
Welding process	No filler material	Typical MIG (filler material is added)
Welding method	Transfer of current through the joined metals	Arc between electrode and the welded metals
Welding equipment	Gun consists of two permanent electrodes	One consumable electrode; arc shielded with inert gas
Control system	Point to point	Continuous path
Drive element	Hydraulic (typical) dc servomotor (seldom)	dc servomotor (typical) Hydraulic (seldom)

Spot welding is a relatively easy process to apply in robotics, and therefore, a substantial number of spot-welding robots are in use on production lines, especially in the automotive industry. The complicated shape of automobile bodies requires spot-welding robots with six axes of motion.

Arc welding usually requires robots with five axes of motion, although there are applications in which four axes are adequate (Engelberger, 1980). Nevertheless, arc welding is one of the most difficult tasks for robots. The electrode must follow a continuous programmed path while maintaining a constant distance from the seam. This requires the matching of four parameters: the arc voltage, the arc current, the rate at which the electrode is fed, and the velocity of the welding gun along the seam. These parameters must be adjusted according to the composition and thickness of the joined metals and the required strength of the seam. Additional difficulty in applying arc welding with robots is that the seam is located slightly differently in each workpiece, and therefore the programmed path does not follow the required one exactly. As a result, a reliable arc-welding robotic system would benefit from on-line sensing devices to trace the path of the actual seam and its distance from the electrode. Appropriate sensors for these tasks (e.g., a vision system or an eddy current sensor to detect the closeness to the workpiece) were under development in the early eighties.

6.4 SPRAY PAINTING

The unhealthy and unpleasant environment of the painting booth in industry made this process an ideal candidate for the application of robots. The solvent materials that are used in spray painting are toxic, and therefore the operators must be protected by masks and be provided with fresh-air ventilation. The painting area must be dust-free and temperature-controlled, and consequently the painting booth is small in size and inconvenient for the operators. Furthermore, the noise arising from the air discharge through the painting nozzles can cause irreversible damage to the ears (Engelberger, 1980). For all these reasons, spray painting became one of the first applications of robots.

6.4.1 Features of Spray-Painting Robots

The requirements for robots in spray painting are different from those of other robot applications, and therefore many robot manufacturers offer a robot dedicated to this one application. Such a spray-painting robot is shown in Fig. 6-12. It has three degrees of freedom in the arm and two in the wrist—pitch and yaw—each with a permissible rotation of 200°. Roll motion at the wrist is not required since the spray gun is a symmetrical tool. This robot can spray only external surfaces. Better

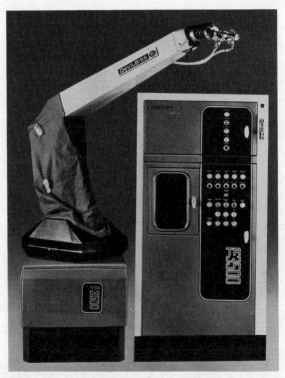

FIG. 6-12   Spray-painting robot (DeVilbiss.)

FIG. 6-13   An "elephant trunk" spray-painting robot produced by Spine Co. and shown at a robotic exhibition in Chicago in 1983.

arm dexterity is required to reach internal surfaces such as the inside of an automobile body.

The spray-painting robots are of CP capability and have the following characteristics: (1) High level of manipulator dexterity, (2) large working volume for small-base manipulator, (3) compact wrist, (4) small payload, (5) low accuracy and repeatability.

Let us elaborate on some of these features. The high mechanical dexterity of the arm is regarded as one of the most important characteristics in the automotive industry. In this industry, the robot needs the ability to paint less-accessible areas such as the inside of the automobile body and trunk, and reach any surface which can be painted by a human operator. To meet these requirements the painting robot arm must be of the articulated type (see Chap. 2) having at least six degrees of freedom.

The automotive industry sometimes uses a snakelike painting robot made of disks and tendons that moves with the agility of an elephant trunk (see Fig. 6-13). The "elephant trunk" enables the robot to go behind obstacles and achieve difficult orientations.

A large working volume and a small robot floor space are usually required in painting because the spray booths are restricted in size. In the automotive industry the average spray booth is 5 m wide (Engelberger, 1980). Therefore, the robot must occupy a small amount of floor space but be able to reach up to the middle of the spray booth (assuming that two robots are working simultaneously in the booth).

The wrist of a spray-painting robot must be compact in order to enable the robot's spray gun to penetrate into narrow spaces. This dictates the removal of the actuators from the wrist through remote drives located on the arm itself. One design approach to reduce the wrist size is the three-roll wrist, which, as the name implies, uses three roll axes to get the pitch, yaw, and roll motions. (The wrist is shown in Fig. 6-11.) The three axes are coincident at one point and are capable of continuous and reversible rotations, features which are important for paint to be deposited on inaccessible surfaces.

The painting robot must be able to carry any type of spray gun. Spray guns, however, are light in weight and therefore painting robots are designed for small payloads (e.g., 1 kg).

Finally, the requirements for repeatability and resolution are the least severe in painting robots. The exact location of end points is not critical, and in many jobs can be even outside the painted surface. Therefore, a repeatability of 2 mm throughout the working volume is regarded as sufficient for spray-painting robots.

6.4.2 Task Programming

Task programming in spray-painting robots is performed by lead-through teaching (see Chap. 7). If the robot is small and light, and the power transmission elements are reversible (i.e., applying a torque at the joint causes the corresponding actuator to move), an expert painter can lead the robot's end effector through the required painting path while the robot is in its teaching mode. Points along the path are recorded and stored in the robot computer as a task program. When the program is played back in the operating mode, the robot nearly emulates the action of the human painter.

When the robot is heavy or contains irreversible elements (e.g., a regular leadscrew) the arm cannot be grasped and a simulator is used. The simulator is a manipulator similar in its dimensions to the robot arm but having no drives and containing extra weights to balance the arm so that an operator can easily move it. The task programming is performed with the aid of the simulator, and the robot can repeat the program upon receiving an appropriate command.

The disadvantage of the lead-through programming method is that if the task program is unsatisfactory it cannot be modified, and the entire program must be recorded again. However, since the same program is used for a large series, as in the automotive industry, the lead-through teaching is not regarded as a substantial drawback and is the most useful programming method for spray painting.

6.5 ASSEMBLY

Assembly systems with industrial robots are mainly used for small products such as electrical switches and small motors. Robotized assembly systems are programmable and therefore provide a cost-effective solution for the assembly of small batch sizes and for batches containing different products. Although industrial robots require the same fixtures, feeders, and other equipment for positioning the parts as conventional assembly machines, simpler workpiece feeders and fixtures may be used because of robots' programmability feature. Furthermore, tactile or optical sensors may be added to the assembly robot to tackle more complex assembly tasks.

6.5.1 Features of Assembly Robots

Fast arm acceleration is very important for assembly robots. In assembly, the traveling paths are relatively short and therefore the joints either move at their maximum speed for short periods only, or they do not reach the speed limit at all. As a consequence, most paths are completed during the acceleration and deceleration periods, and increasing their permissible value shortens the assembly cycle time significantly.

When PTP robots are used in assembly and short paths between end points exist, the assembly time can be readily calculated. Assuming that each joint moves with a triangular velocity profile (i.e., the acceleration is equal to the deceleration and the velocity limit is never reached), the trajectory time t can be determined. The angular distance traveled θ is

$$\theta = 2 \left[\frac{1}{2} \ddot{\theta}_0 \left(\frac{t}{2} \right)^2 \right] = \frac{1}{4} \ddot{\theta}_0 t^2 \qquad (6\text{-}2)$$

where the joint accelerates at the maximum permissible acceleration $\ddot{\theta}_0$ during $t/2$, and then decelerates during $t/2$. Solving for the time t yields

$$t = 2 \sqrt{\frac{\theta}{\ddot{\theta}_0}} \qquad (6\text{-}3)$$

The time is calculated for each of the joints participating in the particular movement, and the longest one determines the trajectory completion time. The sum of these times is the time required to complete the assembly task.

Some assembly tasks require the participation of more than one robot. In order to reduce the cost per arm, there are systems in which

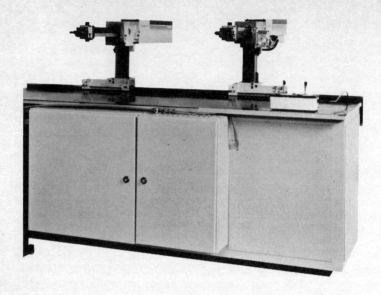

FIG. 6-14 Assembly station consists of two cartesian robots moving along the same base. (General Electric.)

several cartesian arms can use the same base and share the same controller. An assembly station which can be fitted with up to four cartesian coordinate arms is shown in Fig. 6-14.

Assembly robots can be designed in any coordinate system, cartesian, cylindrical, spherical, or articulated. However, many tasks require only vertical assembly motions, such as the assembly of printed circuit boards. For these applications the four-axis robot shown in Fig. 6-15

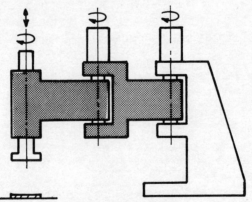

FIG. 6-15 The Japanese SCARA-type robot for assembly in the vertical direction.

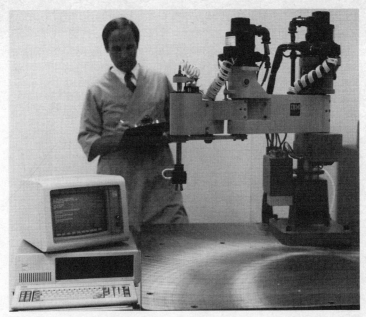

FIG. 6-16   A commercial SCARA-type robot programmed with the aid of a personal computer. (IBM.)

can be adequate. Its arm has two articulated motions, and the wrist has two axes of motion: a linear vertical displacement and a roll movement. This robot can pick up parts located on the horizontal plane, bring them to the assembly location, orient them with the roll motion of the wrist, and finally insert them in a vertical motion.

This class of robot, known as the *selective-compliance-assembly robot-arm-* (SCARA-) type robot, was developed in Japan as a result of an industry-university effort. Five Japanese companies funded research at Yamamachi University to develop a SCARA. There are at present more than ten SCARA-type robots sold in Japan and few in the United States. A table-size SCARA class robot offered in the United States and Europe is shown in Fig. 6-16. It is programmed with the aid of a personal computer, using a subset of the AML programming language (see Sec. 7.3).

6.5.2 Design for Automatic Assembly

The share of human workpower involved in assembly and the number of assembly operators are much higher than in most other manufacturing processes. Therefore, assembly offers the largest potential market for robotics. However, there are many basic difficulties in applying robots to assembly. The main problem is that the parts to be assembled must be prepared, with the right orientation, in special magazines or feeders. In many applications, once this has been achieved, conventional automated assembly systems often become more cost effective than robots. The development of inexpensive intelligent robots equipped with vision can avoid the part preparation stage and increase the number of robots in assembly.

Another problem is that dimensional tolerances of the components can make joining of mating parts very difficult. For example, in order to insert a peg (pin) into a hole, both the repeatability and the resolution of the robot must be less than the clearance. The repeatability can be compensated by a gripper equipped with a force sensor which searches for the hole, but the robot BRU must always be smaller than the tolerance between the mating parts.

Some other problems which have limited the use of assembly robots are

1. Components are often fragile or have a shape which makes them difficult to grip by robots.

2. Human operators routinely perform visual inspection and subsequent small repair tasks on the assembled parts. Similarly, the robot must either repair or reject bad parts, but this is an extremely difficult task for present robot systems.

3. In complex assembly, simultaneous and accurate motion of two hands is required. Such coordination is still difficult for robots.

4. The integration of assembly robots necessitates rebalancing of the assembly line.

In order to respond to these problems, new product design guidelines have recently been introduced, whereby the design is adapted to automatic assembly by robots. The designer must pay attention at the design stage to the joining, handling, feeding, and orientation of parts. The proposed design guidelines are (Boothroyd, 1980; Eversheim & Muller, 1982; Pham, 1982; Waznecke & Walter, 1982):

1. Reduce the number of parts which must be joined together. This means that the product should have fewer parts, but individual parts would be more complex. This rule transfers complexity from the assembly process to the manufacturing process. The need for such complex parts would require more extensive use of modern CNC machine tools.

2. Replace screwed connections by locking connections. Screwing requires numerous individual steps (ordering of screw, setting up, feeding, locating, start threading, and tightening) and a high positional accuracy of the assembly parts. Hence automation of the screwing process requires a considerable effort and a high part quality. Screwed connections should be replaced by simple, self-centering lock connections.

3. Join by motion along a single axis. This provides accurate positioning in cartesian coordinate robots (Fig. 6-14) and in robots having a structure as shown in Fig. 6-15 (assuming that a vertical feeding movement is applied).

4. To facilitate the insertion of mating parts, add guiding slopes and chamfers to holes and cylindrical components (pins, shafts, studs, pistons, etc.).

5. Design part surfaces that can be gripped by robots. For example, parallel surfaces are easily gripped by a parallel jaw hand. Note that if the part becomes more complex (rule 1 above), its geometry may introduce difficulties in handling by robots.

6. Eliminate components which are difficult to be gripped or be fed (bendable components such as hoses and cables), or improve their form. For example, springs tend to hook to each other. By closely rolling their winding ends this problem can be eliminated.

7. Design symmetric parts which facilitate the feeding and the orientation in the assembly. If a part is completely symmetrical and reversible, it needs no orientation.

8. In view of anticipated developments in vision sensors, consideration should be paid to the optical characteristics of components, such as color, surface finish, and light reflectivity.

9. Reduce the complexity of final assembly by defining new subassembly groups. In the final assembly, parts are often difficult to handle because of a complex contour or an untoleranced surface. Therefore, where possible, many assembly tasks should be transferred from the final assembly stage to subassembly stages.

The problem of product design for assembly by robots has generally been recognized, but until it penetrates into the design office, a lengthy learning process will be necessary. Designers need to become aware of the consequences of even the smallest design changes on the effort needed for automatic assembly.

The designer must have an idea about the structure of the assembly station. A typical assembly station will comprise one or more industrial robots, component feeders, and a conveying device. Visual and force sensors can be added, enabling the system to compensate for positioning errors of the robot, deviations in part dimensions, and orientation of the components. In order to achieve efficient design for automatic assembly, the designer must know at the outset the types of robots, feeders, sensors, and grippers used in the station. This information, properly digested and applied, will greatly enhance the utility of robots in industrial assembly.

6.6 MACHINING

There are five basic types of machine tools to perform machining: drilling machine, lathe or turning machine, milling machine, shaper, and grinder. Out of all these machining operations, only drilling is being successfully done with robots, and mainly in the aircraft industry. Another application related to machining which is performed by robots is deburring metal parts. Most metal parts made by machining operations (either by machine tools or by mass production machines) contain burrs, that is, rough edges or ridges left on the machined surfaces. The removal of these burrs can be done by robots.

6.6.1 Drilling

In the production of the F-16 military aircraft, a group of four wing skins are attached to the wing understructure by fasteners. For these fasteners holes have to be drilled. Although the holes vary in size for the different types of fasteners, over 6000 holes are made using only a few different sizes (Kusmiersky, 1979). The required tolerances are within ±0.25 mm (0.01 in), a number which dictates the repeatability of the appropriate robot. The method used to drill these holes before the robotic age has been to put template holes over the wing and use manual drilling.

Robots can replace the manual operators if the template hole is provided with a chamfered guide. The gripper of the robot holds a portable pneumatic drill and guides it from hole to hole. At each hole, a fixed drill cycle is performed, and then the robot moves the drill to the next hole.

Programming the robot to perform the task is quite simple. Since drilling is a PTP operation, the manual teaching method (see Chap. 7) is appropriate. The programming and control methods are much more complicated when CP machining operations (e.g., deburring) are applied.

6.6.2 Deburring Metal Parts

Burrs are generated almost always when machining is performed on metal parts. Burrs can be generated between a machined surface and a raw surface, or at the intersection between two machined surfaces. The removal of these burrs is an expensive operation. Most deburring is performed manually by workers equipped with appropriate tools. By closely following the manual method, the industrial robot can solve most deburring problems.

There are two basic ways to perform robotized deburring. If the part is relatively lightweight, it can be picked up by the robot and brought to the deburring tool. If the part is heavy, then the robot holds the tool as shown in Fig. 6-17. The support of the tool is very important, whether it is held by the robot or mounted on the work table. In both cases the relative motion between the tool and the part is of a CP type with high repeatability (approximately 0.2 mm) and highly controlled speed (Weichbrodt, 1979). Therefore, deburring is one of the most difficult tasks for robots.

Robots for deburring operations are either programmable servo-controlled robots or adaptive robots. In the first type, the robot is programmed to move along a predetermined path. Vibrations of the tool

FIG. 6-17   Deburring operation performed by a robot. (ASEA.)

are avoided by adding damping elements between the tool and the tool holder.

Adaptive robots use sensors to detect the size of the burrs in real time. They can be of two types: force sensors and surface sensors. Force sensors can measure from one force component up to six parameters—three forces and three torques. The one-component force sensor is used in industry to compensate for tool wear in real time or to measure heights of burrs. The appropriate corrections are then inserted into the task program data.

The surface sensors are capable of measuring burr geometry in real time. The sensor is in contact with the surface and is driven along it by the robot arm, while its output is used to adjust the position of the tool. At present, surface sensors as well as multicomponent force sensors are at the research stage and are not applied in practice.

REFERENCES

Bjorke, O.: "Computer-Aided Part Manufacturing," *Comp. Ind.*, vol. 1, 1979, p. 39.

Boothroyd, G.: *Design for Assembly Handbook*, Univ. of Mass., Amherst, 1980.

Datsko, J.: *Material Properties and Manufacturing Processes*, J. Datsko Consultant, Ann Arbor, Mich., 1979.

Engelberger, J.: *Robotics in Practice*, AMACOM, New York, 1980.

Eversheim, W., and W. Muller: "Assembly Oriented Design," *3rd Int. Conf. Assemb. Automat.*, May 1982, pp. 177–190.

Hitomi, K., M. Nakajima, and Y. Osaka: "Analysis of the Flow-Type Manufacturing Systems Using the Cycle Queuing Theory," *Trans. ASME, J. Eng. Ind.*, vol. 100, November 1978, p. 468.

Jablonowski, J.: "Robots That Weld," *Am. Mach.*, Special Report 753, vol. 127, no. 4, April 1983, pp. 113–128.

Koren, Y.: *Computer Control of Manufacturing Systems*, McGraw-Hill Book Co., New York, 1983.

Kusmiersky, T.: "Robot Applications in Aerospace Batch Manufacturing," *Ind. Robots*, vol. 2, SME, 1979, pp. 169–182.

Merchant, M. E.: "The Factory of the Future—Technological Aspects," *Towards the Factory of the Future*, ASME, vol. PED-1, November 1980, pp. 71–82.

Pham, D. T.: "On Designing Components for Automatic Assembly," *3rd Int. Conf. Assem. Autom.*, May 1982, pp. 205–214.

Stauffer, R. N.: "Flexible Manufacturing System—Bendix Builds a Big One," *Manuf. Eng.*, vol. 87, August 1981, pp. 92–93. (a)

Stauffer, R. N.: "Automating for Greater Gains in Productivity," *Manuf. Eng.*, vol. 87, November 1981, pp. 58–59. (b)

Tanner, W. R.: *Industrial Robots*, vol. 2, Applications, Society of Manufacturing Engineers, Dearborn, Michigan, 1979.

Warnecke, H. J., and J. Walter: "Automatic Assembly—State-of-the-Art," *3rd Int. Conf. Assem. Autom.*, May 1982, pp. 1–14.

Weichbrodt, B., and L. Beckman: "Deburring of Metal Parts in Production," *Ind. Robots*, vol. 2, SME, 1979, pp. 161–168.

Wu, S. M., and D. S. Ermer: "Maximum Profit as the Criterion in Determination of the Optimum Cutting Conditions," *Trans. ASME J. Eng. Ind.*, November 1966, pp. 435–440.

CHAPTER 7

Programming

The combination of six axes of motion and a noncartesian coordinate system in which most robots operate make the task programming of a robotic system much more difficult than the programming of other manufacturing systems, such as the part programming of NC machine tools. Therefore, many robot systems apply on-line programming methods in which the robot itself is used at the programming stage. The on-line programming approaches are referred to as *teaching-by-showing* methods, since during the programming stage the operator moves the robot arm through a set of required points, or a path in space, by means of push buttons, a control handle, or a joystick, and the system remembers the points or path so that the robot can reproduce these motions during execution. Off-line methods are characterized by programs that tell the robot what to do, but the robot itself is not used during the programming stage. When applying off-line programming, the software provided by the robot manufacturer should be powerful enough to enable the programmer to think in terms of tool coordinates. This software makes the calculations needed to manipulate the robot joints in order to achieve a certain location and orientation of the end effector.

In teaching or programming a task, the robot system is provided with the following information:

1. Coordinates of significant points. These points can be divided into three major categories: *end points*, which are points that the end effector must accurately reach at the end of each trajectory; *via*

points, which define a path through which the end effector passes in order to avoid obstacles; and *reference points*, which determine what path the arm should take under specified conditions. Conditions can be represented by the status of external signals from sensors, or by the magnitude of internal numeric values.

2. Gripper status at each point, i.e., closed or open.

3. Velocity values for each motion.

4. A definition of the sequence of operations or the cycle of operations.

Three programming methods are used in robotics: manual teaching, lead-through teaching, and programming languages. The first two are on-line programming methods while the last one is an off-line method. These methods are discussed below.

7.1 MANUAL TEACHING

This type of programming is the simplest and the most frequently used in PTP robotic systems. Teaching is done by moving each axis of the robot manually until the combination of all axial positions yields the desired position of the robot. The commands of these manual motions are given by the operator, who uses a series of push buttons on a control box, or teach pendant, as shown in Fig. 7-1. When the desired position is reached, the operator stores the coordinates of this point into the computer memory. This process is repeated for each required position until the task program is completed.

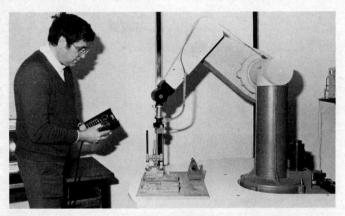

FIG. 7-1   The use of a teach box in manual programming.

A typical manual control box is shown in Fig. 7-2. In order to provide teaching flexibility, several teach modes are available: joint, world, tool, and free. With each of these modes, the position of the robot arm can be manipulated. This is done with the aid of six toggle switches (i.e., three-position switches with the center position spring loaded).

For each of the first three teaching modes (i.e., joint, world, and tool), the desired motion speed can be selected by using the speed knob. In order to enable precise positioning of the joint, an incremental motion (INC) is provided, in which each joint can move an amount corresponding to one pulse of the position feedback device (e.g., incremental encoder) each time the joint toggle switch is pushed. When the robot arm is operated, the speed is determined by instructions written into the task program by the programmer.

The joint mode allows control of individual joints in teaching. Each joint can be moved by its corresponding toggle switch. For example by flipping switch 1 to the left, the base axis (joint 1) will rotate to the left (clockwise); or by flipping switch 6 to the right, the joint next to the tool (joint 6) rotates to the right (counterclockwise).

The joint mode is not convenient for task programming, and either the world or tool modes is preferable. In the world mode the robot's TCP moves in a reference coordinate system fixed in the robot arm base. For example, if the first switch is flipped to the left, the TCP is moving along the $-X$ direction. In order to accomplish this motion, the robot computer calculates in real time the required individual joint motions. The switches marked RX, RY, and RZ allow motion around the X, Y, and Z axes, respectively.

In the tool mode the motion is controlled in the same manner as in the world mode, but now the tool moves in its own coordinate system. That means that a cartesian coordinate system is attached to the tool, with its origin at the center of its mounting flange. A $+X$ instruction moves the tool

FIG. 7-2 The teach box of the PUMA robot.

along the positive direction of the X axis of this coordinate system. Similarly, RX rotates the tool around this X axis, etc. Note that in the tool mode, RZ is a motion of joint 6 alone.

In addition to these three teaching modes, a free mode is also available. This mode enables a limited lead-through teaching (see Sec. 7.2). In this mode, flipping a toggle switch releases the corresponding joint; namely, its control becomes disabled, its brakes are released, and the joint becomes completely free swinging. The joint can now be positioned by the operator at any desired point.

When a desired point has been reached with any of the teaching modes, the point can be stored in the memory of the robot computer by pressing the RECORD button. Storing a point means recording the position of each joint including the gripper (i.e., open or closed, can be selected on the teach box).

Finally, in order to return control to the robot computer, a COMP button is available. The COMP mode is selected when control of the robot arm by a task program is desired or for arm calibration. Arm calibration is achieved by sending each joint to the closest position which coincides with an index pulse from the corresponding encoder (see Chap. 2).

A teach box which includes a wider variety of manual control functions is shown in Fig. 7-3. This is the teach box of the five-axis robot shown in Fig. 2-23. In addition to the control options available on the teach box in Fig. 7-2, it also includes a selection of the type of each recorded point, a coarse selection of the *operating* speed, and a checking mode of the program during teaching.

Three types of points can be stored:

1. An *end point,* which is a regular point at the end of a trajectory. In the operating mode, end points should be reached accurately, and therefore the arm motion is instantaneously stopped at these points.

2. A *via point,* which is used when the arm has to move beyond obstacles. Passing through the exact coordinates of a via point is not important, and therefore the arm motion does not stop at these points.

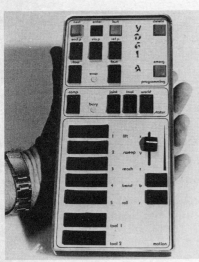

FIG. 7-3   The teach box of the Technion's robot.

3. A *reference point* is an end point which contains a conditional branch instruction. For example, if a sensor transmits a specific signal the arm must move to a certain position, and if the signal is different, the arm should move to a different position.

In the operating mode the arm moves at a certain speed unless instructed differently for a specific trajectory. In practice, two operating speeds are adequate for many applications: a fast speed for coarse motion and a slow speed when approaching a target very accurately, as in assembly of two mating parts. Accordingly, the teach box permits the programming of two operating speeds (fast and slow) in the teaching mode. The values of these two speeds can be selected for each task program. If another speed is desired for a certain trajectory, then the programmer must alter the corresponding speed in the program.

Finally, the teach box permits the checking of the program, point by point, by using the next and previous point buttons (NEXT and LAST). During the check a point can be deleted or modified as desired.

The points stored in the teaching mode are the geometric basis of the task program. In the operating mode, the path between two successive points depends on the type of the controller: PTP or CP (see Chap. 2). In simple PTP robots, each axis moves at its maximum velocity until it reaches the required coordinate. Consequently, some axes will reach their required coordinate before others, and the path of the robot end effector between the points is unpredictable. When the stored points are fractions of an inch apart, PTP robots can be used in CP applications.

The method which uses manual teaching in CP applications applies a linear interpolation between the points. While in a PTP system the path between successive points is unpredictable, with the linear interpolator feature this path is a straight line, as shown in Fig. 7-4. During the programming stage the points can be selected so as to attain the required accuracy. Another way to use the manual method in CP applications is to paste an adhesive strip along the required path on which small and equal segments are marked. To assist the programmer the segments can be numbered as shown in Fig. 7-5. This method is used in

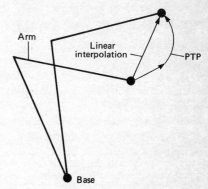

FIG. 7-4 The difference between the path with PTP control and with linear interpolation.

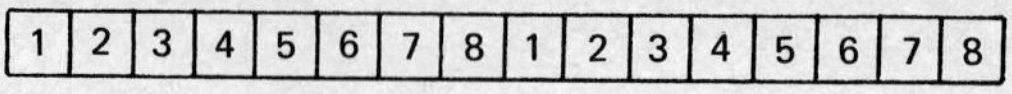

| 1 | 2 | 3 | 4 | 5 | 6 | 7 | 8 | 1 | 2 | 3 | 4 | 5 | 6 | 7 | 8 |

FIG. 7-5   An adhesive strip divided into small segments to be used in manual teaching of CP applications.

programming of arc-welding robots. With the aid of the manual controller the programmer moves the welding gun from one point to the next and records each point in the memory of the robot computer. The recorded program is then played back and additional points are added if necessary. Finally the programmer specifies the robot velocity along the path, which depends on the other welding parameters (see Sec. 6.2).

The main disadvantage of manual teaching is that the equipment is tied up in the programming stage. However, because of its simplicity, this is the method most acceptable to industry, especially in PTP applications such as machine loading and unloading, simple assembly tasks, and spot welding.

7.2 LEAD-THROUGH TEACHING

One might think that the simplest method of programming CP robots is by physically grasping the robot end effector and leading it through the desired path at the required speed, while simultaneously recording the continuous position of each axis. There are robots in which this method can be applied (by disengaging the motors with electric robots or reducing the oil pressure in hydraulic robots). However, because of the transmission elements (such as gears and leadscrews) in the robot manipulator, it might be impossible to generate a motion of the robot joints by pulling its end effector. One solution is to construct another identical manipulator, equipped with position feedback devices (usually encoders), but with no drives and transmission elements attached. This device is denoted as the *robot simulator* or *teaching arm*, and is illustrated in Fig. 7-6. To facilitate the grasping of the simulator, counterbalanced weights are sometimes added. The simulator is manually grasped by the operator and led through the required path and at the same time the position of each axis is sampled at a constant frequency and stored in the computer. Once taught, the robot is capable of repeating the operation as many times as required.

While the advantage of the lead-through teaching method is the simplicity of direct programming, it has some major disadvantages:

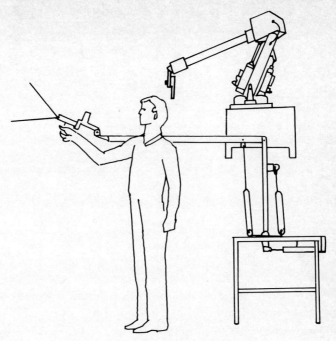

FIG. 7-6 A robot simulator for lead-through teaching.

1. An investment in a simulator is required. Alternatively, if the robot itself is used in a teaching mode, then the robot is tied up during programming, which consequently reduces its overall efficiency.

2. Every operator motion is recorded and played back in the same manner. Therefore, unintentional motions will also be played back, unless the system allows the removal of unwanted moves. In practice, many trials may be required before achieving the final program.

3. Since teaching is performed manually, a high precision in generating paths cannot be achieved.

4. In the programming stage, it is impossible to obtain the exact required velocity along the path. Other methods for programming the velocity must be applied.

5. A considerable memory size is required to store the data. As an example, assume that the positions of six axes are recorded during 5 min at a rate of 50 samples per second. The amount of required memory words is $6 \times 5 \times 60 \times 50 = 90,000$. For this reason, disk storage devices are usually used with this method.

Lead-through teaching is widely used in spray painting of auto parts or other products.

7.3 PROGRAMMING LANGUAGES

Programming languages in robotics are concerned with the generation of all data required to move the robot end effector along a required path in order to perform a specific task. While the on-line manual teaching procedure is sufficient for many applications, it can become very tedious when hundreds of points must be programmed individually on PTP or CP arms. Such a condition frequently exists in the aerospace or automotive manufacturing industries. For example, in aerospace there are hundreds of holes that must be drilled in sheet metal parts and many rivets that must be properly placed to hold the metal parts together. In the automotive industry, hundreds of spot welds must be applied to each car body to ensure proper assembly (Tarvin, 1980). In the vast majority of cases, the coordinates of all the point locations involved in such applications have already been defined in drawings relative to some part reference frame. Furthermore, this information frequently exists in data bases within the customer's plant. It is indeed wasteful and redundant for a robot operator to program such points manually using current on-line teaching techniques. In cases like these, an off-line approach to programming, wherein data base information can be used, seems to offer many advantages.

The commercially available robot languages in use today provide instructions to move the manipulator, read sensors, send output signals, and many other instructions that simplify task programming. Using these languages, the programmer instructs the end effector in English-like commands and is not concerned with the motion of the individual joints. These motions are calculated by the language processor, which can be either a compiler or an interpreter. Motion instructions in programming languages contain the following types of data:

1. Positional information—the start and end points of each trajectory
2. Trajectory type—linear (in CP operation) or quickest path (in PTP mode)
3. Velocity
4. Functions to be performed at a point, such as a time delay, tool manipulation, and sending or receiving signals
5. Gripper status—closed or open.

The sequence of operations required to perform the task is dictated by the task programmer.

The main problem in off-line programming is the positional errors that are caused by the deflection of the manipulator links. The deflections depend on the payload and position of the end effector and are not taken into account by the language processor. This is particularly important when dealing with large arms designed to lift heavy loads. Therefore, in many robot systems programming is divided into two stages: off-line and on-line. In the off-line stage end points are assigned identifying names (e.g., POINT A). In the on-line stage the loaded manipulator, used as a digitizer, is brought to each end point and the actual point coordinates are recorded for each identifying name. The drawback of this method is that the manipulator is occupied during one stage of programming.

A number of new robot programming languages have been recently developed, such as AL at Stanford University, PAL at Purdue University, RPL at SRI International, SIGLA at Olivetti, TEACH at Bendix, HELP at DEA (now G.E.'s Allegro), RAIL at Automatics, and MCL (contract of the U.S. Air Force ICAM program) (Wood and Fugelso, 1983). An example of an off-line programming language is AL, developed for use with a robot system which consists of two light-payload arms that can interact while performing assembly tasks. Language statements consist of motion instructions such as

MOVE ARM TO (position)
OPEN HAND TO (value)

conditional statements such as

ON FORCE (value) DO (operation)

and other statements defining boolean expressions, assigning values to variables, etc. (Mojitaba and Goodman, 1979; Tarvin, 1980).

One of the best-known commercial languages is VAL, which is used with the PUMA arms of Unimation Inc. (Motiwalla, 1982; Shimano, 1979; Unimation Inc., 1977). Programs in VAL are written on the same computer (DEC LSI-11) that controls the robot. A VAL program contains two parts: (1) coordinates of end points and (2) motion and action statements. The point coordinates can be determined by manually leading the manipulator to each required point and recording its coordinates. The motion section of VAL consists of a series of one-word commands that are followed by information such as

> MOVE (location), means a PTP motion command
> MOVES (location), means a straight-line interpolation command

Each MOVE command calls a subroutine that calculates the individual joint positions required during the move to the new location.

The coordinate system is established by FRAME, and can be shifted with the SHIFT command. The gripper action is specified with OPEN and CLOSE, and the velocity is given in SPEED statements.

To demonstrate the VAL language, let us assume that the robot must pick objects from a chute and place them in successive boxes. One possible sequence of robot activity is as follows (Unimation Inc., 1977):

1. Move to a location above the part in the chute.
2. Move to the part.
3. Close the gripper jaws.
4. Remove the part from the chute.
5. Carry the part to a location above the box.
6. Put the part into the box.
7. Open the gripper jaws.
8. Withdraw from the box.

The corresponding VAL program is as follows:

```
• PROGRAM DEMO.1
    1.   APPRO PART,50
    2.   MOVES PART
    3.   CLOSEI
    4.   DEPARTS 150
    5.   APPROS BOX,200
    6.   MOVE BOX
    7.   OPENI
    8.   DEPART 75
```

The exact meaning of each line is:

1. Approach PART, i.e., move to a location 50 mm above the part in the chute.
2. Move along a straight line to the part.
3. Close the gripper jaws; the "I" means a short delay before executing the next statement.
4. Withdraw the part to a point 150 mm above the chute along a straight-line path.

5. Move along a straight line to a location 200 mm above the box.

6. Move the part into the box.

7. Open the gripper jaws.

8. Withdraw to a point 75 mm above the box.

When the program is executed, it causes the robot to perform the steps which describe the task.

Another commercial robot language is AML (A Manufacturing Language) which was developed at IBM (Grossman, 1982). AML is a general computer programming language first and a robot language second. AML provides arithmetic operations (e.g., addition), boolean operations (e.g., AND, OR, etc.), relational operations (e.g., LE for ≤), conditional expressions (IF . . . THEN . . . ELSE), and looping expressions (WHILE . . . DO and REPEAT . . . UNTIL). For controlling the robot arm AML includes a set of commands such as MOVE, DMOVE, GUIDE, SPEED, and MONITOR.

To demonstrate the AML language let us assume that the robot has to take 10 blocks from a feeder and palletize them. The appropriate AML program is as follows:

```
EXAMPLE: SUBR;
  I: NEW 0;
     WHILE (I = I + 1) LE 10 DO              -BEGIN LOOP
     BEGIN
                                             -TO FEEDER
     DMOVE(Z,9.0);                           -UP BY 9 INCHES
     MOVE(XY, ⟨3.525, 1.078⟩);               -TO FEEDER
     MOVE(Z,1);                              -DOWN TO BLOCK
     IF SENSE(IRBEAM) EQ 0            -CHECK INFRARED BEAM
         THEN BEGIN
             TYPE('BLOCK MISSING', EOL);
             RETURN; END;
                                             -PICK UP BLOCK
     GRASP;                          -CALL SUBROUTINE TO GRASP
     MOVE(Z,5);                               -MOVE UP
                                             -TO PALLET
     MOVE(XY, ⟨14.021, 2.0o1-11.304⟩);
                                             -TO PALLET
     MOVE(Z,1);                               -DOWN
     RELEASE;              -CALL SUBROUTINE TO RELEASE BLOCK
     END;                      -END OF WHILE . . . DO LOOP
     END;                      -END OF EXAMPLE SUBROUTINE
```

The programming language is actually a tool with which a human can describe his or her understanding of the required task to the robot. The level of the language is considered higher if the description of the task is simpler and involves less programming effort. At the lowest level are the *joint model* languages (Ardayfio and Pottinger, 1982), in which the programmer programs the motions of the individual joints. This language level does not exist in commercial robots.

In most commercially available robot languages, such as VAL and AML, the programmer programs the position of the end effector and is not concerned with the motion of the individual joints. Therefore these languages are categorized as *end-effector level* (Kempf, 1982) or *manipulator model* languages (Ardayfio and Pottinger, 1982).

There are higher-level languages in which motions are described in terms of the object being manipulated and accordingly are denoted as *object level* (Kempf, 1982) or *world model* (Ardayfio and Pottinger, 1982; Heginbotham et al., 1979; Mojitaba and Goodman, 1979) or *model-based* (Grossman, 1982) languages. Programming languages in this category utilize world models, namely symbolic representations of the objects located in the robot environment. Languages in this category include AL, PAL, RAPT (University of Edinburgh) (Poppleston et al., 1978) and Autopass (IBM) (Lieberman and Welsey, 1977). These programming languages understand the geometry of the robot and its environment and therefore can generate collision-free instructions to control the robot in space. Autopass (*auto*matic *p*arts *a*ssembly *s*ystem) is an experimental world model language developed by IBM for assembly tasks (Grossman, 1982; Lieberman and Welsey, 1977). Autopass can execute the two following statements:

```
PLACE BRACKET IN FIXTURE SUCH THAT
    BRACKET.BASE CONTACTS FIXTURE.TOP

PLACE INTERLOCK ON BRACKET SUCH THAT
    INTERLOCK.BASE CONTACTS BRACKET.TOP
    AND INTERLOCK.HOLE IS ALIGNED WITH
    BRACKET.HOLE
```

World model programming systems (including Autopass) are experimental and are not currently used in industry.

In the far future there might be even higher-level languages. With these languages the programmer will only have to define the goal to be achieved. For example, ASSEMBLE A CARBURETOR. The programming system must have access to preestablished data bases and enough knowledge in assembly techniques to translate this statement into a sequence of instructions understood by the robot control system.

The advantage of using programming languages over the teaching by showing (manual and lead-through) methods is that the robot is free during the major portion of the programming stage. In addition, these languages allow flexibility when corrections in the task program are required. Nevertheless, the use of programming languages has several drawbacks:

1. *Large gap in required skills.* The use of programming languages is difficult for workers and managers in the manufacturing industry. On the other hand, many programmers do not understand the problems on the shop floor. This lack of communication results in many problems in using programming languages in industry (Nagel, 1984).

2. *No standardization in languages.* Different manufacturers use different languages. If a robot user purchases equipment from several manufacturers, then several different languages must be learned.

3. *Positioning errors.* On many robots the actual position of the end effector depends on the particular manipulator and on the load. Different loads cause different deflections of the links in the manipulator. When manual teaching is used, the robot is taught with the actual manipulator and load and therefore in the playback stage positional errors due to these factors are not introduced. Programming languages do not take the robot loads into account, which causes uncertainty in the positioning of the robot.

4. While the use of languages frees the robot in the first programming stage, the equipment is still tied up in the program debugging stage, which might be time consuming when complex tasks are programmed (Tarvin, 1980).

7.4 PROGRAMMING WITH GRAPHICS

Graphics can be added to the programming system in order to emulate the robot motions by performing an animation of the robot on the graphic terminal linked to the computer. Figure 7-7 shows a robot dis-

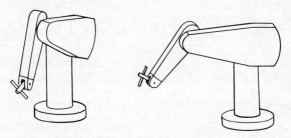

FIG. 7-7 A solid model emulation of a robot, produced by Synthavision. (CDC.).

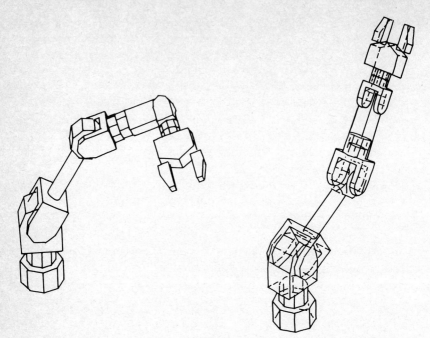

FIG. 7-8   A wire-frame diagram of a robot produced at the University of Tokyo.

played in two positions. To produce an animated effect, the moving model needs to be displayed at a number of intervals between the programmed positions. The environment of the robot (e.g., the work table, feeders, obstacles, etc.) is displayed as well. The user is provided with facilities to change various aspects of the model interactively. This can include robot motions, grip and release, types of conveyor movement (free-running or indexed), time delay, etc. (Heginbotham et al., 1979). The user can check if the robot could manipulate objects to various positions without exceeding arm or wrist constraints and without hitting objects in the work space.

The model of the robot can be presented either as a solid model with the removal of hidden lines (see Fig. 7-7), or as a "wire-frame" model, as shown in Fig. 7-8. The wire-frame models are less useful for many manufacturing needs, such as describing the changing status of a block of metal during machining, but in robotics their drawbacks are less significant and they can be used.

In addition to the advantages provided by languages, programming with graphics has the following advantages:

1. Graphical debugging of the task program, which reduces the time and effort in debugging the programs on the real robot.

2. The user can start experimental programming before the robot is purchased. The user can also test different robots on the screen and select the most appropriate one for the planned application.

3. The robot environment can be viewed while emulating the movements of the robot in order to avoid collisions. This is demonstrated in Fig. 7-9, where an automobile body is displayed together with a spot-welding robot.

4. The computer-aided design and computer-aided manufacturing (CAD/CAM) common data base of the factory of the future can be utilized. For example, if the shape of an automobile body is stored in the CAD/CAM computer of an automobile manufacturer, the

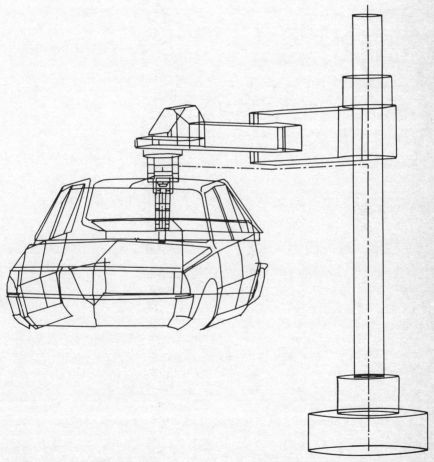

FIG. 7-9 Graphic display of a robot carrying out spot welding on an automobile body. (Renault.)

designer can specify on the screen the points which should be spot-welded. These points will be transferred via a computer network to the control computer of the spot-welding robot.

Despite of all these attractive features, programming with graphics has not been used in industry in the early eighties. The major drawbacks of this method are the following:

1. The computer terminal and the associated software are very expensive (more than $100,000 in 1982). The user can purchase an extra robot for this price and execute the task program on the real equipment.
2. There is a wide gap in skills between the people who write the task programs and the people who use them on the shop floor.
3. There are no standards in equipment supplied by CAD/CAM vendors and robot manufacturers. The user expends effort in learning many systems, which becomes inefficient and costly.

7.5 STORING AND OPERATING TASK PROGRAMS

The two most common ways to write a task program are either by a teach box or with the aid of a programming language. With both methods programs are stored in the task program section of the robot computer. In principle, the robot task can be taught in the JCS, WCS, or in the TCS.

7.5.1 Point-to-Point Robots

In PTP robots only the JCS is used. In the teaching mode, with the aid of the teach box, the joints are directly manipulated until reaching a required point and then the positions of each joint are stored in the JCS. Similarly, when a programming language is used the end-point joint positions are stored. At the beginning of each segment, the stored position values of the joints are directly used to command the control loops.

7.5.2 Continuous-Path Robots

Kinematic programs are used for simulation of robot motions in programming with graphics. However, the main use of kinematics is in the algorithms: (1) The direct kinematics, which is used to solve for the

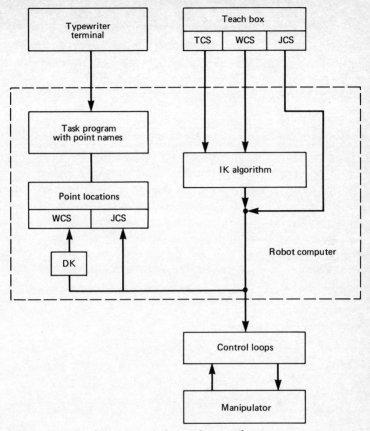

FIG. 7-10   CP robot system in the teaching mode.

motion of the end effector in the robot base coordinates on the basis of a given motion at each joint, or in other words, determining the position in the WCS from given values in the JCS. (2) The inverse kinematics, which is used to calculate the individual joint motions on the basis of a given motion of the end effector, or in other words, determining the JCS values from a given position in the WCS. The inverse kinematics is the heart of the interpolator algorithm contained in the control program of a CP robot (see Chap. 5). It is obvious that for PTP robots there is need for neither direct nor inverse kinematics since both the control program and the task program utilize only joint values.

In CP robots task teaching can be performed in the JCS, WCS, and TCS, as shown in Fig. 7-10. Programming with the JCS, however, is inconvenient and consumes much programming time, and therefore the WCS or the TCS is preferable. When operating the teach box in the

WCS, the desired positions and orientations of the tool are directly fed to the inverse kinematics (IK) routine of the interpolator (see Fig. 7-10), which calculates in real time the reference commands for the control loops. When a desired end point is reached by the programmer, either the corresponding joint values are read and stored in the JCS, or the desired joint values can be retransformed by the direct kinematics (DK) block and stored in the WCS.

When the TCS is applied, the procedure is similar to that of using the WCS. In the teaching mode, however, additional algorithms are required. Some of these algorithms perform coordinate transformations from the TCS to the WCS, using direct kinematics techniques.

In the operating mode, if the point locations of the task program have been stored in the WCS, they are directly used as input to the interpolator at the beginning of each new trajectory (see Fig. 2-25 in Chap. 2). If the point locations have been stored in the JCS, a direct kinematics calculation from the JCS to the WCS must be performed at the beginning of each new trajectory. The interpolator adds new reference points along a straight line connecting the previous and the new end points (see Fig. 2-6), and subsequently retransforms the locations to JCS values.

It is worthwhile to note that another case where direct kinematics is needed is when information given by sensors is used as a feedback signal to the control program through a decision-making algorithm. The sensor can identify the location of an object in either the WCS, TCS, or in another coordinate system, and this information must be transformed to the coordinate system in which the control program is operating. For example, the sensor can be mounted on one of the manipulator links (see Fig. 1-3), which consequently requires a transformation from the links' coordinate system to the WCS.

REFERENCES

Ardayfio, D. D., and H. J. Pottinger: "Computer Control of Robotic Manipulators," *Mech. Eng.*, vol. 104, no. 8, August 1982, pp. 40–45.

Grossman, D. D.: "Robotic Software," *Mech. Eng.*, vol. 104, no. 8, August 1982, pp. 46–47.

Gruver, W. A., B. I. Soroka, J. J. Craig, and T. L. Turner: "Evaluation of Commercially Available Robot Programming Languages," *13th Int. Symp. Ind. Robots & Robots 7*, vol. 2, April 1983, Chap. 12, pp. 58–68.

Heginbotham, W. B., M. Dooner, and K. Case: "Assessing Robot Performance with Interactive Computer Graphics," *Robot. Today*, Winter 1979–80, pp. 33–35.

Kempf, K. G.: "Robot Command Languages and Artificial Intelligence," *Proc. Robots VI*, 1982, pp. 369–391.

Lieberman, L. I., and M. A. Welsey: "AUTOPASS—An Automatic Programming System for Computer Controlled Mechanical Assembly," *IBM J. Res. Dev.*, vol. 21, no. 4, 1977, pp. 321–333.

Mojitaba, S., and R. Goodman: *AL User's Manual,* Stanf. Artif. Intell. Lab., Memo AIM-323, January 1979.

Motiwalla, S.: "Development of a Software System for Industrial Robot," *Mech. Eng.* vol. 104, no. 8, August 1982, pp. 36–39.

Nagel, R. N.: "Robotics: A State of the Art Review," *The 4th Jerusalem Conf. on Inf. Tech.*, Jerusalem, May 1984, pp. 566–577.

Poppleston, R. J., A. P. Ambler, and I. M. Bellos: "RAPT—A Language for Describing Assemblies," *Ind. Robot*, vol. 5, no. 3, 1978, pp. 131–133.

Shimano, B. E.: "VAL—A Versatile Robot Programming and Control System," *Proc. IEEE 3rd Int. Comput. Software Appl. Conf.* Chicago, 1979, pp. 878–883.

Tarvin, R. L.: "Considerations for Off-line Programming a Heavy Duty Industrial Robot," *10th Int. Symp. Ind. Robots*, Milan, Italy, March 1980, pp. 109–114. (Also published in *Robot. Today*, Summer 1981, pp. 30–35.)

Unimation Inc.: *Users Guide to VAL,* October 1977.

Volz, R. A., T. N. Mudge, and D. A. Gal: "Using ADA as a Robot System Programming Language," *13th Int. Symp. Ind. Robots & Robots 7*, vol. 2, April 1983, Chap. 12, pp. 42–57.

Will, P. M., and D. D. Grossman: "An Experimental System for Computer Controlled Mechanical Assembly," *IEEE Trans. Comput.*, vol. C-24, no. 9, 1975.

Wood, B. O., and M. A. Fugelso: "MCL, The Manufacturing Control Language," *13th Int. Symp. Ind. Robots & Robots 7*, vol. 2, April 1983, Chap. 12, pp. 84–96.

CHAPTER 8

Sensors and Intelligent Robots

The common feature of all robots which have been discussed so far is their ability to repeat a preprogrammed sequence of operations as long as necessary. However, these robots are unable to sense and respond to any change in their environment. For example, if a robot was programmed to grip a part at a certain point, the robot will always close its gripper jaws at that point, even if a part is not there. If an obstacle is inserted in the robot's path, the robot will collide with it and will not move around the obstacle to avoid collision.

In order for robots to operate effectively in a changing environment, they must be equipped with sensors and have some degree of artificial intelligence (AI). Based upon the data received from the sensors, the AI algorithm makes real-time decisions that might change the programmed sequence of operation of the robot. Such robots are classified as intelligent robots.

A block diagram of an intelligent robot is shown in Fig. 8-1. It consists of a two-level heirarchical closed-loop control system. At the lower level there are the position control loops of the individual joints; each includes a position feedback device, such as an encoder. In many robots, a velocity control loop (with a tachometer as the velocity-measuring device) is contained at the low control level as well.

The higher level in the heirarchy contains a sensor that is able to sense the robot environment, its associated interface, and the AI algorithm. The loop is closed through the interpolator algorithm, which

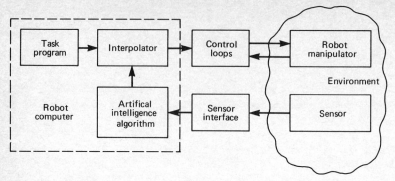

FIG. 8-1   Block diagram of intelligent robot system.

responds to the original task program instructions with corrections obtained from the AI algorithm.

8.1 INTRODUCTION TO ROBOTIC SENSORS

The sensors most likely to be useful in robotics may be classified as contact and noncontact. Contact sensors may be further subdivided into tactile sensors and force-torque sensors. A tactile or touch sensor indicates a physical contact between the end effector carrying the sensor and another object. A simple tactile sensor is a microswitch. Tactile sensors are used to stop the motion of a robot when its end effector makes contact with an object. This can be used to avoid collision or to signal the robot system that a target was reached or to measure object dimensions in inspection.

Force and torque sensors are located between the gripper and the last joint of the wrist, or on a load-bearing member of the manipulator, where they can measure reaction forces and moments. They employ either piezoelectric transducers or strain gauges cemented onto the compliant sections. Force and torque sensors are further discussed in Sec. 8.5.

Noncontact sensors in robotics include proximity sensors, visual sensors, acoustic sensors, and range detectors. A proximity sensor senses and indicates the presence of an object within a fixed space near the sensor. For example, eddy-current sensors can be used to precisely maintain a constant distance from a steel plate. One of the simplest robotic proximity sensors consists of a light-emitting diode (LED) transmitter and a photodiode receiver that senses light when a reflecting surface is near. The main drawback of this sensor stems from the dependency of the received signal on the reflectance of the intruding object.

Other proximity sensors are based on capacitive and inductive principles.

The visual sensing systems are more sophisticated and are usually based on television cameras or laser beam scanners. The signals of the cameras are preprocessed in hardware and can be fed at a rate of 30 or 60 frames per second into a computer. The computer analyzes the data and extracts the required information, such as the presence, identity, position, and orientation of an object to be manipulated, or part integrity and completeness of products under inspection. Visual sensing systems are further discussed in Sec. 8.2.

An acoustic sensor senses acoustic waves and interprets them. The level of sophistication of acoustic sensors varies from a primitive detection of the presence of acoustic waves to recognition of isolated words in continuous human speech. In addition to human-robot voice communication, acoustic sensing can be utilized by robots to assist in controlling arc welding, to stop the motion of a robot when a loud crash is sensed, to predict a mechanical breakage about to happen, and to inspect objects for internal defects (Nitzan, 1981).

Finally, there is the class of noncontact systems that uses a projector plus an imaging device to obtain surface shape information or range information (see Sec. 8.3). Other range detectors are based on sonar or an ultrasonic device.

There are two basic methods of using sensors: static sensing and closed-loop sensing. Frequently, sensors are used in robot systems in a mode wherein sensing and manipulation alternate. That is, the manipulator is stationary while sensing is being done, then motion is done without further reference to the sensors. This method is referred to as *static sensing*. With this method, the vision is used to establish the position and orientation of an object to be grasped, then the robot moves blindly to the spot.

By contrast, in closed-loop sensing, the robot is controlled by the sensing device during the manipulator motion. Most vision systems operate in a closed-loop mode, where the vision system monitors the error between the robot's actual position and the desired position. This error is used to actuate the robot drives. With closed-loop sensing, even if the object is in motion, for example, on a conveyor, the robot is able to grip it and transfer it to a desired location.

However, a number of factors have prevented much progress in closed-loop sensing in the early eighties. A primary difficulty is that the length of time necessary to analyze and image is almost as long as the time it takes the robot to move from one position to another. To be useful, visual analysis times should be short enough to allow several pictures to be taken and interpreted during an arm motion.

When applying force and tactile sensors during the manipulation motion, the response time is not as much of a problem as it is for visual sensors because existing sensors deliver much sparser information, i.e., six forces and torques at the wrist, or a low-resolution array of binary sensors in the fingertips. As sensors become more sophisticated, we can expect that more information processing will be necessary to make use of the sensory data.

8.2 VISION SYSTEMS

Most vision systems are equipped with one or more video cameras linked to a vision processor. The vision processor digitizes the camera image and analyzes it to define the object.

The main applications of vision systems are in handling, assembly, part classification, and inspection. In handling and assembly the vision systems are used to recognize the position and orientation of objects to be handled or assembled. These systems can also determine the presence or absence of parts and detect particular features of the object (e.g., diameter) in order to screen out parts which do not meet specifications.

Vision systems are also used for identification of objects for part classification, where, in most cases, the parts are passing on a conveyor belt through a robotic station.

Vision systems are being used, with and without robots, in a variety of inspection tasks. Product inspection is a major area in manufacturing, where a significant increase in productivity is assured when using automated visual inspection systems. In addition, these systems provide better quality control, since human inspectors find it difficult to concentrate throughout the work shift and errors in their judgment become more likely.

The simplest types of vision systems are able to identify isolated parts and their orientation against a flat surface. No overlapping objects are permitted, and the number of stable positions of each object is limited. Constant lighting conditions must be maintained so as to clearly define the silhouette of a dark part on a white background (or vice versa) for binary image processing. Diffuse backlighting will usually give excellent results. Diffuse front lighting may prove satisfactory if the objects contrast with the background (e.g., gray objects against a black conveyor or vice versa). Stray lighting, such as typically found in an industrial environment, must be shielded.

A robot interfaced with a vision system is shown in Fig. 8-2. Two round parts are placed on a flat surface and are shown on the television

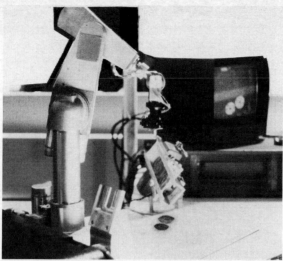

FIG. 8-2   A robot vision system, demonstrated at Robots VI exhibition in Dearborn, Michigan, in 1980.

monitor. The location of one part is detected by the vision system, which instructs the robot to pick it and place it at the top of a sloped track. As the robot places the part, another one is released and falls on the surface in a random location. Notice that in this demonstration the parts are completely symmetric with one stable position, which facilitates the task of the vision system.

This and other robot vision systems are capable of picking up isolated parts from a flat surface, or even from an organized pile. The problem of picking parts from a disordered pile (known as the bin-picking problem), however, is a more difficult one but can be solved by a multistep operation. For example, a group at the University of Rhode Island has developed a three-step method to acquire and orient overlapping parts in a bin (Birk et al., 1979, 1980; Kelley et al., 1979). First, they approximate the location of any one object. Second, using sensory feedback from a vacuum gripper, they make the robot pick up the object and transport it to a flat surface. Finally, using a nonoverlapping-objects algorithm, they orient the part and transport it to its final destination. Notice, however, that the algorithm will not necessarily acquire the topmost object. In fact, the initial acquisition by the robot is completely at random. While this may be sufficient for some applications, other applications require the recognition of overlapping objects in their overlapping state.

The problem of picking randomly oriented parts from a bin could be solved if vision capability were available to analyze the jumble of parts.

The intensive research in vision around the world guarantees that improved visual processing capabilities will shortly allow robots to work with overlapping objects on cluttered backgrounds with unconstrained conditions.

8.2.1 Vision Equipment

Most vision systems in robotics consist of a video camera interfaced with the vision computer through a video buffer called the *frame grabber* as shown in Fig. 8-3. The camera's analog video signal can be sent to a conventional television monitor which continuously presents the scene to be scanned. When instructed, a frozen frame of the camera is scanned row by row, and the corresponding output signal is stored in a digital form in the frame grabber. The digitized image is subsequently transferred from the frame grabber to the vision computer in parallel bytes for image processing.

Employing the vision computer to perform the entire image processing results in a slow system which is impractical for most applications. Therefore, most vision systems also contain a hardware video preprocessor to perform selective data reduction and initial processing, as shown in Fig. 8-3.

In many vision systems, the images of the objects received from the cameras are reduced to their silhouettes. This silhouetting is done as the part is scanned by the television camera, using a technique called *thresholding* as shown in Fig. 8-3. The system assumes that objects of interest will contrast sharply with the background; thus all object inten-

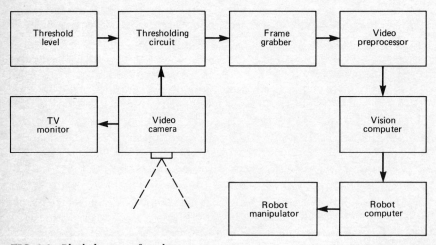

FIG. 8-3 Block diagram of a robotic vision system.

sities will be above or below a certain level. This threshold value is used to determine which intensities should be set to white and which to black during silhouette formation.

Cameras. There are three types of cameras which are typically used in vision systems: linear arrays, vidicons, and solid-state cameras.

Linear Arrays. A linear array of photosensitive elements can be used where the scanned scene is in continuous constant motion, as for example on a moving conveyor. The camera scans a line across the conveyor, and the motion of the conveyor produces the other dimension of the picture. Linear arrays were available in 1983 with resolutions of up to 2048 elements.

Vidicons. The vidicon camera is the usual device for television signal generation and is frequently used in vision systems. Vidicons are inexpensive due to their high-volume production for entertainment purposes. However, their limited tube life and fragility may cause problems when used in manufacturing environments.

Solid-State Cameras. In many industrial applications, cameras containing arrays of photosensitive elements are used. These cameras utilize either charge-injection device (CID) or charge-coupled device (CCD) technology. Solid-state cameras were available in 1983 with resolutions of 256×256 picture elements, or pixels, but higher resolutions can be expected.

Mounting Cameras. There are two basic approaches for mounting cameras in vision systems: mount it on the robot arm (Agin, 1979), or mount it externally (Brady, 1981).

When the camera is mounted on the arm, the robot is moved based upon an error signal derived from the object's location within the field of view. Such a robot system is used at SRI International to control the robot hand as it picks an object from a noninstrumented moving conveyor belt. The system can obtain samples at a rate of ½ Hz (Rosen & Nitzan, 1977). A disadvantage of this type of system is that it requires a camera to be mounted on the robot where it is subject to rough handling. The camera also increases the robot arm loading and requires running extra cables to the robot.

An alternative method of implementing visual feedback is to mount the camera externally and use it to observe the relative position between the robot end effector and objects in the work space. The error between the observed and desired positions is used to control the robot. This type of visual feedback, however, is dependent upon knowledge about the environment to determine three-dimensional information from a single view. If the television camera is mounted on the robot arm, different views of the object can be observed by changing the location of the arm, and three-dimensional information can be obtained.

8.2.2 Image Processing

The digitized image is stored in the memory of the vision computer and subsequently processed. The computer can determine the type of object and its orientation and transfer this information to the robot computer. Basically, the image of the object can be presented and stored in two forms: as a binary image or as a gray-scale image (Binford, 1982; Kanade, 1983).

Binary and Gray-Scale Images.   The digitized image consists of pixels; the brightness of each is represented in the computer by a number. In a binary vision system, the brightness is represented either by 0 or 1, corresponding to black and white. In a gray-scale system, the brightness can take on a range of values. For example, if 4 bits are reserved for each pixel, the gray-scale values are from 0 (black) through 15 (white).

Binary data are generated by the use of a thresholding circuit, which may be fed by the video output of the camera as shown in Fig. 8-3, or by a digitized output produced by an analog-to-digital converter (ADC). Many binary vision systems can automatically select the optimum threshold level for a particular application. In other systems the threshold level is adjusted manually. Gray-scale systems require an ADC to generate the data for the frame grabber.

Cost-effective vision systems are using binary image processing, but the use of these systems places restrictions on the scene to be scanned. The information to be extracted from the image must be apparent from silhouettes, and therefore the depth of the object cannot be detected. Gray-scaled systems require more hardware and computing resources, but they permit the detection of a surface curvature and the ability to distinguish between overlapping parts.

Binary Processing.   In a binary image system a two-color map of the visual scene is stored in the computer, with 1's representing the silhouette of the part and 0's representing the background.

The hardware video preprocessor (shown in Fig. 8-3) is used in binary vision systems to improve the signal-to-noise ratio and to reduce the amount of information sent to the computer. One of the most common tasks of the hardware preprocessor is edge detection. Edge detection is based on the fact that there is often a difference in brightness

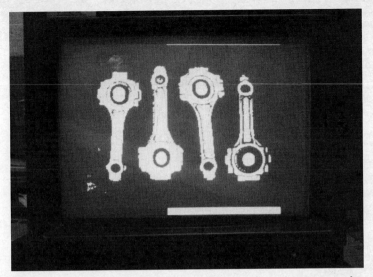

FIG. 8-4   A binary image of a part is displayed on the vision system in the Center for Robotics and Integrated Manufacturing at the University of Michigan.

between object and background. By detecting the transitions from 0 (black) to 1 (white) and back again at the thresholding circuit output, the edges of a white silhouette on a black background are decoded. The video preprocessor sends only the transition points to the computer, thereby reducing the amount of information stored in the computer. The computer software contains an edge-detection algorithm which determines what edge the point belongs to by examining neighboring points, then entering the point's location in a list for that edge. If the point is isolated, however, the system assumes a new silhouette and creates a new list.

The size of the neighborhood or "window" examined by the edge-detector algorithm determines its sensitivity to noise. The presence of a noise spike in the pixel array can lead to the detection of false edges. To help avoid false edges, adjacent pixels can be averaged by the edge-detection algorithm to reduce the effect of the noise. A survey of edge-detection techniques can be found in Brady (1981).

Once the image is represented in memory, the task of enhancement and analysis can begin. Image enhancement usually refers to the processing performed to improve the quality of a picture for analysis. An enhanced binary image of a part is shown in Fig. 8-4. Vision analysis in robotics usually refers to part recognition and determination of its position and orientation.

Part Recognition. For many robot vision systems used in handling and assembly, only the location and orientation of objects is needed. In these cases, the vision system computes the location and orientation of the silhouettes. A silhouette's location is defined as its geometric center of gravity and its orientation is defined as the orientation of an ellipse that has the same area. The location and the orientation are transmitted from the vision computer to the robot controller and the object is subsequently picked up by the robot arm.

For part classification and for some handling applications, the recognition of objects is required. In these cases, geometric properties of the pictured object may be measured and compared to stored models of known objects. Models may take many forms, ranging from stored images of parts to statistical properties computed from the edges of objects, or even geometric relationships between dimensions on the object.

One straightforward recognition method is based upon comparison of the image of an object with prestored images in order to find the best matching. However, in most cases the object will have a different orientation from that of its prestored image. For matching, the pictured image may have to be rotated to the orientation of the prestored image, which is a long computational process. Therefore, in practice, this method is restricted to cases in which objects might enter the vision scene with a limited number of orientations. Examples of such applications include the recognition of printed characters or parts placed at a certain orientation on a conveyor belt.

A useful part recognition method is based upon comparison of geometric features of the parts. Common features used for the comparison test are:

The total area (including holes) or the net area

First moments of area for calculating the centroid

Maximum and/or minimum radius of the image

Perimeter length

Perimeter squared divided by the area

Length and width of the bounding rectangle

Number of holes and their area

In most cases a single distinguishing test is not sufficient and a set of features must be compared in order to recognize the part.

As an example, consider the two parts shown in Fig. 8-5. Part A is a square with a side of 2 cm, and part B is a 2.5-cm-diameter disk with a

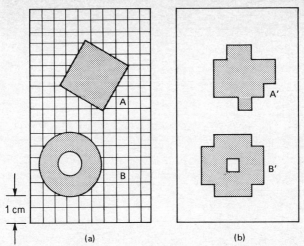

FIG. 8-5 (*a*) Parts and (*b*) their digitized images.

0.9-cm-diameter hole. A set of four features is selected to compare the parts: The maximum radius, the perimeter, the net area, and the hole area. The measurements are taken from Fig. 8-5 and given in Table 8-1.

From the values in Table 8-1 it becomes clear that the parts cannot be distinguished by the maximum radius and the perimeter of the silhouettes. They can be distinguished, however, by comparing the net area and the hole area. Many times it is easier to recognize a part by the number of blobs. A square with two holes, for example, represents three blobs: the square and the two holes. The number of blobs in flat parts does not depend on the position of the part, and this facilitates the identification process.

This and other image-processing algorithms are based on matching features of the pictured silhouette with features of models stored in the computer memory. The closeness of a match is determined by scoring

TABLE 8-1 Measurements of the Parts in Fig. 8-5

Type	Part	Maximum radius	Perimeter	Net area	Hole area
Real part	A	1.41	8.00	4.00	0
	B	1.25	7.85	4.27	0.69
Silhouette image	A′	1.50	10.00	4.25	0
	B′	1.50	10.00	5.00	0.25

individual feature matches and then weighting and combining the individual scores to create a total score.

8.3 RANGE DETECTORS

To perform many factory tasks, such as assembly or picking parts out of bins, two-dimensional vision systems are not sufficient, and a measurement of a distance should be provided as well. A range sensor measures the distance from a reference point, usually located on the sensor itself, to a point in the scene. Range sensors that are used with robots can be classified into four types:

1. Range finding based on a laser scanner or a supersonic or radar sensor (Nitzan, Brain, and Duda, 1977; Will and Grossman, 1975).

2. An optical sensor that can determine distance based on the amount of reflected light (Okada, 1982; Rosen and Nitzan, 1977); this sensor uses active illumination denoted as structured light.

3. Visual sensor with active illumination operating on a triangular principle (Agin, 1979; Agin and Binford, 1973; Nitzan, 1981; Okada, 1982; Thompson, 1981).

4. Stereo vision system with passive illumination (ambient light) (Brady, 1981; Duda and Hart, 1973; Thompson, 1981), also operating on a triangular principle.

The first two methods, namely the range-finding and reflected-light techniques, are easiest to implement and thus closest to commercial realization. Both techniques rely on the way light or sound waves or x-rays reflect from the surface of objects for depth measurement. Range-finding systems typically measure the reflection time of a laser beam to the object and back again to measure its distance.

The Triangular Principle.   A visual short-range sensor which projects light on an object is shown in Fig. 8-6. The sensor consists of a light source and an array of light-sensitive elements. These photosensors are scanned to detect the one having the maximum output signal. Assuming that the ith element has the maximum output, and its distance from the light source is a_i, there are two homologous triangles which yield the following relationship

$$\frac{x}{x+h} = \frac{b}{a_i} \qquad (8\text{-}1)$$

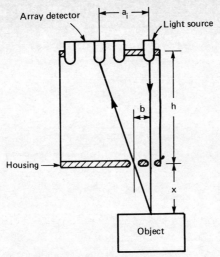

FIG. 8-6 The triangular principle implemented in an optical range sensor.

The distance x between the sensor and the object is obtained from Eq. (8-1):

$$x = \frac{bh}{a_i - b} \qquad (8\text{-}2)$$

The resolution of this sensor is proportional to the distances between the photosensor elements. However, if the distances are too small, the signal processing for detecting the range becomes more difficult, since there may be more light-sensed elements.

Three-dimensional vision systems that rely on range finding or active (structured) light are inherently limited because they require interaction with the object under observation. Although these systems may be adequate for many practical applications, a vision system should be based on passive illumination to avoid putting constraints on the observed objects or their environment. One such system is based on a stereo vision technique.

Stereo Vision. Stereo, or binocular, vision uses two television cameras to obtain a view of a scene, similar to the two eyes of a human being. The slight disparities in the images received by the left and right eyes enable humans to determine the shape and relative depth of the visible surfaces. Similarly, stereo vision deals with matching two descriptions computed from the separately processed images.

A major obstacle in the development of stereo vision systems is what is known as the "correspondence" problem: objects in each view must be matched with one another before the disparity between the two views can be determined. Matching is a problem because, as a result of parallax, an object appears in slightly different places in the right and left views and may be partially or totally obscured in one view.

Most work on automating stereo vision chooses a succession of small windows (typically 10 by 10) from one image, and in each case searches for the window in the other image that optimizes the correlation between their gray levels (Brady, 1981). A shortcoming of current correlation techniques, however, is that their accuracy is limited to a fraction of the window size (typically 5 pixels).

Despite intensive research efforts, range sensing has so far (1984) hardly been utilized in performing robotic tasks. More research effort is needed to improve range sensor capabilities (higher speed, higher resolution, higher accuracy, wider range, and smaller size) and to reduce the cost of such sensors.

8.4 ASSEMBLY-AID DEVICES

The assembly of mating parts by an industrial robot requires high precision. High-precision robots, however, are expensive and usually fit only a very limited range of applications. In order to use a standard industrial robot in assembly it has to be equipped with some sort of feedback for correction of inaccuracies. In general, there are two types of suitable devices: a *passive* one, such as the remote-center compliance, and an *active* force-torque sensor in which contact forces are measured and fed back for correction. Both types are mainly used in assembly tasks as an aid in the insertion of round pegs into holes, usually with chamfered corners, as shown in Fig. 8-7, and in the insertion of screws.

Both the passive and the active devices compensate for the positional inaccuracy of the end of the last link of the robot manipulator. However, while with the passive accommodation devices (e.g.,

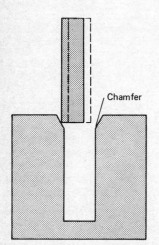

FIG. 8-7   Peg–hole insertion in assembly.

Chamfer

remote-center compliance) the position of the manipulator does not change, during the stage of correcting the peg's positional error, the active devices (e.g., force sensor) employ a search strategy according to which the manipulator position varies in order to find the center of the hole.

8.4.1 Peg-in-Hole and Screw Insertions

In order to examine how frequently peg-in-hole and screw insertions exist in assembly, a study of ten products was made (Nevins and Whitney, 1978, 1980). Eight of these products were of cast or machined metal, one of molded plastic, and one of a variety of plastic, sheet-metal stamps, and wires. The latter product was the exception to the findings from the others: 70 percent of the parts arrive from one direction, 35 percent of all tasks are single peg–hole insertions, and screw insertions represent 25 percent of all tasks. Notice that since most parts arrive from one direction, it should be possible to assemble them with devices which have much fewer than six degrees of freedom (Nevins & Whitney, 1978), and consequently there are cases in which only three-component force sensors are successfully used (Borenstein, 1983; Salmon, 1976; Spalding, 1982). This study justifies the efforts made in developing force and torque wrist sensors as well as their counterpart passive devices, such as the remote-center compliance and other mechanisms which are attached to the last link of the manipulator as assembly-aid devices.

8.4.2 The Remote-Center Compliance

In typical assembly the peg is pivoted from the top and touches the chamfered corner of the hole, as shown in Fig. 8-7. In order to continue the insertion, the peg must be slightly rotated and pushed until a two-point contact is achieved. A further insertion is possible only if the top of the peg can move laterally (i.e., in the horizontal plane in this case).

This sequence of operations is automatically performed by a linkage device called a remote-center compliance (RCC) which was developed at Draper Laboratory (Nevins and Whitney, 1978, 1980). The basic structure of the RCC is shown in Fig. 8-8. The RCC is attached to the last link of the robot and holds the peg to be inserted. The RCC consists of a series of linkages connected with flexible elements, where the remote center is located at point O, at the end of the peg. One set of linkages allows the peg to rotate about its tip if it is angularly misaligned with the hole. A second linkage cascaded with the first allows the peg to move from side to side to correct a lateral error without introducing

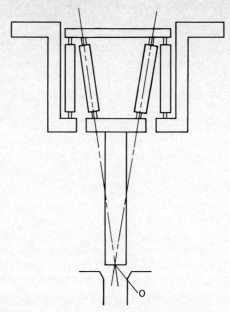

FIG. 8-8 The remote-center compliance (RCC) developed at Draper Laboratory.

unwanted rotation. The combination of independent lateral motion and rotation about the tip makes the RCC a useful tool in assembly.

There are reports (Nevins and Whitney, 1978) that the RCC performs difficult assembly tasks, such as putting a bearing into a housing with a clearance ratio of 0.0004 in ⅛ s starting from a lateral error of 1 mm and an angular error of 1.5°. The clearance ratio is defined as follows:

$$\text{Clearance ratio} = \frac{\text{diameter of hole} - \text{diameter of peg}}{\text{diameter of hole}}$$

While the RCC is a simple, cost-effective device with fast response time and does not need any adjustments (as force sensors do), it has several drawbacks:

1. A particular RCC can be used with a limited range of peg lengths (for example, 40 ± 5 mm). If the assembled product contains pegs with wide variations in length, the RCC device must be frequently changed and its usefulness becomes questionable.
2. The RCC is inappropriate for assembly of pegs in the horizontal direction. Usually the insertion angle must be much less than 45°.
3. The RCC does not work properly if the chamfered corner of the peg

does not initially contact the chamfered corner of the hole. This limits the size of positional errors that the RCC can accommodate (typically below 1.5 mm). The RCC cannot be used in chamferless insertion tasks.

None of these drawbacks exist for force sensors. Force and torque sensors, however, are more expensive; they require calibrations to compensate for drifts in the associated electronic circuits; and their peg-insertion time is longer, since they employ hole-searching algorithms which require computation time.

8.5 FORCE AND TORQUE SENSORS

Force and torque sensors measure three components of the force and three components of the torque acting between the gripper and the manipulated object. In most cases the transducer is located between the last link of the robot and its gripper (Borenstein, 1983; Nitzan, 1981; Van Brussell & Simons, 1978) as shown in Fig. 8-9. Such a transducer is denoted as a *wrist force sensor*, and it measures the components of the force and torque by transducing the deflections of the sensor's compliant sections.

Force sensors in robotics employ either piezoelectric transducers or strain gauges cemented onto the compliant sections. A prime requirement for a wrist sensor is high resolution, which is obtained by high signal-to-noise ratio and low hysteresis. These two features depend on the quality of the sensing elements, the mechanical structure of the sensor, and the design of the associated electronics.

Application of force and torque sensors to industrial robots has not been practicable in the early eighties for two reasons (Nitzan, 1981). The ability of robots developed during the seventies to respond to low-amplitude, high-frequency motion commands has been poor because of the low bandwidth of robot control systems. This has been one obstacle because force sensors usually have a very high gain from displace-

FIG. 8-9 A force sensor is located between the robot's last link and its gripper. (The sensor was built in the robotics laboratory, Technion.)

ment to force, requiring the robot to move very short distances to maintain desired forces at the tool. The new generation of robots, designed with more attention to high performance, appear to be able to respond at the required bandwidth.

The second obstacle to application of force sensors to industrial robots has been the limited real-time adaptability of robot programming and control systems. Early systems had very limited ability to respond to real-time inputs, usually only the ability to alter the execution sequence of program steps. In order to perform force-controlled motions, the control system must be able to modify individual robot motions in real time. This ability is just now appearing in the latest programming and control systems.

Features which must be considered when designing force sensors are

1. The sensor should be as small and as light as possible in order not to limit the wrist dexterity and the payload of the robot.

2. The sensor must be able to sense forces up to the limit of the robot. This maximum force, together with the force resolution (in grams), dictates the number of stages of the ADC in the computer interface.

3. The sensor should have overload protection to avoid damage in collisions.

4. A decision regarding the number of sensed force-torque components (i.e., three or six), should be made at an early design stage.

The last point is further discussed below.

8.5.1 Six-Component Sensors

A six-axis wrist sensor based on strain gauges and elastically flexing beams, built at SRI International, is shown in Fig. 8-10 (Nitzan et al., 1977). Similar six-axis sensors were designed and built at Stanford University (Binford, 1973), at Draper Laboratory (Nevins and Whitney, 1978), at the Catholic University of Leuven (Belgium) (Van Brussell and Simons, 1978), at the IPA (West Germany) (Warnecke, Schweizer, and Haaf, 1980), and at other research institutes.

The SRI sensor, as well as many other force sensors, uses strain gauges cemented on elastic beams. The strain gauges are selected because of their low cost, high resolution, and reliability. The elastic beams in the SRI's sensor were designed to avoid hysteresis and to separate the force and torque components mechanically at the earliest possible point in the sensing scheme. As seen in Fig. 8-10, four beams are oriented with their long axes in the Z direction (labeled P_{x+}, P_{y+}, P_{x-}, and P_{y-}), and four are perpendicular to the Z direction (labeled Q_{x+}, Q_{y+}, Q_{x-}, and Q_{y-}). One end of each beam has a neck. Because of its

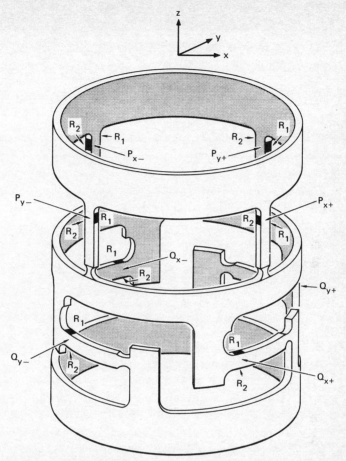

FIG. 8-10 Six-component wrist sensor based on strain-gauge elements. (SRI International.)

small dimensions, this neck transmits negligible bending torque and thus increases the strain at the end where the gauges are located.

As shown in Fig. 8-10, two foil strain gauges, denoted by R_1 and R_2, are cemented on each of the eight beams such that a vector from the center of R_2 to the center of R_1 points along the positive direction of either one of the X, Y, and Z axes. These gauges are oriented to respond to tensile or compressive strains along the long axis of the beam. On beams P_{x+} and P_{x-}, the gauges are placed on the two faces perpendicular to the Y direction. Likewise, on beams P_{y+} and P_{y-}, the gages are placed on the faces perpendicular to the X direction. On beams Q_{x+}, Q_{x-}, Q_{y+}, and Q_{y-}, the gauges are placed on faces perpendicular to the

Z direction. Each gauge is placed as near as possible to the end of its beam that has no neck. The two strain gauges R_1 and R_2 on each beam are connected in a potentiometer circuit, the output voltage of which is denoted by the name of its beam. For example, P_{x+} is the output voltage of the pair of gauges on beam P_{x+}.

When a peg–hole insertion task is applied, the voltages proportional to the three force and three torque components acting on the wrist are obtained by the following linear combinations:

$$F_x \propto P_{y+} + P_{y-}$$

$$F_y \propto P_{x+} + P_{x-}$$

$$F_z \propto Q_{x+} + Q_{x-} + Q_{y+} + Q_{y-} \tag{8-3}$$

$$M_x \propto Q_{y+} - Q_{y-}$$

$$M_y \propto -Q_{x+} + Q_{x-}$$

$$M_z \propto P_{x+} - P_{x-} - P_{y+} + P_{y-}$$

These six components of the force and torque are subsequently sent to the computer for processing. The computer, in turn, generates six correction signals which are sent as references to the six corresponding joints. The manipulator will move the peg according to these signals toward the center of the hole in order to insert it into the hole.

Nevertheless, if a wrist sensor were to operate only in the described way, it would hardly be useful. In fact, the system could find the required position but could do nothing to hit the right position once missed. In order to accomplish the system, it must have a certain compliance between the grasped peg and the end of the robot's wrist. When employing passive devices, such as the RCC, this compliance is provided by the device itself. With active devices, the mechanical flexibility of the wrist plays the same role. It is also worthwhile to note at this point that drilling with robots is possible only because of the flexibility of the wrist, since the drill is self-aligned through a guide hole in a template, and this guidance actually provides the fine positioning of the drill.

8.5.2 Three-Component Force Sensors

Most six-component force-torque sensors were developed in research laboratories. Three-component force sensors are produced by robot manufacturers (Salmon, 1976; Spalding, 1982), but are still hardly used in industry. The preference given by robot manufacturers to three-component force sensors may be attributed to the following reasons:

1. The cost associated with three-component sensors is considerably below that associated with six-component systems (Spalding, 1982).

2. Six-component sensors require reading of many transducers simultaneously, which represents a problem when computations are performed in real time (Salmon, 1976).

3. The measurement of the torques is unnecessary since the force application point is always known (Salmon, 1976).

4. It seems that torque about the tool axis does not need to be known with high precision and can probably be determined adequately by sensing the drive current to the motor of the tool-rotation axis of the robot (Spalding, 1982).

The limitation of a three-component force sensor is its inability to perform insertion operations in any direction. In addition, this sensor can locate a peg (inserted from the top) at the center of the hole but cannot aid in pushing it down into the hole.

A three-component force sensor was designed for the PUMA-600 robot in the Robotics Laboratory of the Technion in Israel (shown in Fig. 8-9). The sensor is 70 mm in length, 51 mm in diameter, and weighs 200 g. The structure of the sensor is shown in Fig. 8-11. It consists of four parts: thin-walled cylinder (A), inner protective cylinder (B), outer protective cylinder (C), and torsional protection pin (D).

Eight strain gauges are located on the thin-walled region of A as shown in Fig. 8-12. As can be seen from Fig. 8-13, strain gauges Z_1 through Z_4 are interconnected as a full bridge, such that the output ΔV_z of the bridge is not affected either by changes in temperature nor by moments about the X or Y axis. Therefore, the Z bridge is sensitive only to axial loads. Strain gauges X_1 and X_2 are interconnected as a half bridge (with the complementing half inside the interface), such that the output ΔV_x is insensitive to changes in temperature, axial loads, or moments about the X axis (i.e., forces in the Y direction). Thus the X bridge is sensitive to forces in the X direction or moments about the Y axis only. The Y bridge operates similarly and is sensitive to forces in the Y direction only.

The thin-walled cylinder is protected from overloads in the following manner: Forces acting on the center of the gripper in the X or Y direction cause an internal force and moment at section b (see Fig. 8-11). This will cause a deflection, and a deflecting angle, at section b. For any force exceeding 50 N, these deflections will cause contact between the protective cylinders B and C at sections b and a, such that B and C will bear any load above 50 N up to 1000 N.

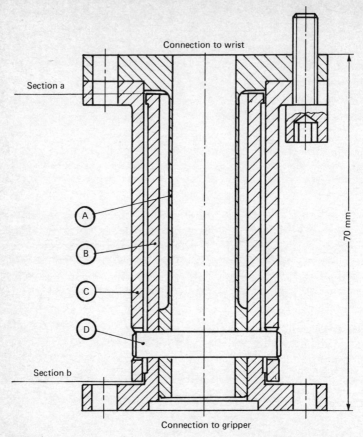

FIG. 8-11 Structure of the Technion's three-component force sensor.

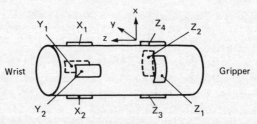

FIG. 8-12 Strain-gauges location on the thin-walled cylinder.

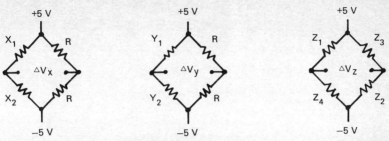

FIG. 8-13   Interconnections of the sensor's bridges.

A block diagram of the entire force-sensing system is shown in Fig. 8-14. The bridge outputs are amplified and filtered through a second-order low-pass filter. The amplified signals are compared with threshold levels, which may be adjusted by potentiometers. If the signal level exceeds the threshold level, a binary signal is sent to the input/output (I/O) module of the PUMA robot.

Even though the bridges are theoretically temperature compensated, the slightest inaccuracy in a gauge location will result in unequal temperature influence on the various gauges of the bridge. If a high sensitivity is required, the sensor must have high gain, which in turn causes considerable drift as a result of temperature variations caused by the heat generated in the strain gauges. In order to eliminate the effect of drift, a *zero unit* (see Fig. 8-14) has been introduced into the system, and it fulfills a dual purpose:

1. It allows redefining of any signal level as 0 V at any time, either manually or by a running robot program. If, for example, the zeroing signal is given at the start of a robot program, then drift signals accumulating up to this time are neglected and only forces at the gripper cause deviation from the newly defined zero level.

2. The zero unit also compensates for different states of the gripper. Since any change of orientation of the gripper influences the force measurements (because of the gripper's dead weight), sending a zeroing instruction just before the force measurement will eliminate the undesired effect of gripper orientation.

An additional feature of this sensor is the *reference unit*. The reference unit receives binary signals from the I/O module of the PUMA and translates them into threshold levels instead of those determined by the potentiometers. Thus the threshold level for each one of the three force channels may be changed independently while running a program as often as desired. One important application of the reference unit

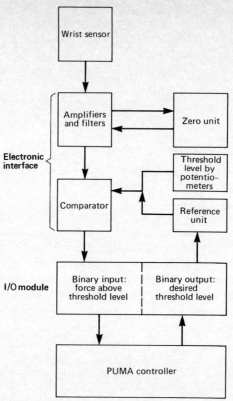

FIG. 8-14 Block diagram of force sensor system.

(besides the obvious need in assembly tasks) is the classification of parts by weight during handling by the robot.

It is worthwhile to note that there are systems in which the gain of the sensor can be set at either of two levels (Spalding, 1982). The lower gain is used when the full load might be applied to the sensor. The higher gain provides the increased resolution required at low force levels. With the high gain, however, drift problems are more severe.

In order to insert a peg in a hole with the Technion's sensor, a search algorithm was programmed on the PUMA computer using VAL instructions (see Chap. 7). The peg is brought down in the Z direction to the programmed position, which is the center of the hole. However, because of arm inaccuracy, many times the peg will miss the hole, and then a contact with the part is made. The resultant reaction force in the Z direction is measured by the sensor and sent to the PUMA controller, which, consequently, immediately stops the approach instruction in the

Z direction and starts the search routine. The search for the hole is performed in a quadratic spiral as shown in Fig. 8-15, where the distance between each two parallel lines is 0.1 mm, which is equal to the robot's BRU. If the resolution is smaller than the clearance (i.e., BRU $< D - d$, where D and d are the hole and peg diameters, respectively), then the peg must be above the hole during the spiral motion. Otherwise the hole must have a chamfered corner with a diameter D', such that BRU $< D' - d$.

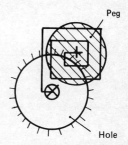

Peg

Hole

FIG. 8-15   Search strategy for the center of a hole.

During the search procedure, the reaction force in the Z direction exists (it varies slightly due to surface roughness). However, when the entire peg is above the hole the Z force suddenly becomes zero and the search stage comes to an end. Upon detecting the change in the Z force, the system returns to the first strategy of pushing the peg in the Z direction. An insertion of several millimeters is performed and then the gripper jaws are opened and the peg falls into the hole.

The system contains also a self-learning routine, by which the location at which the peg was inserted is stored and used as a target point for the next peg insertion. This can save searching time in cases where the inaccuracy is caused by a systematic error.

Figure 8-16 shows an assembly task in which the search strategy is applied. This task consists of inserting two long pins and screwing them from the side with the aid of an electric screwdriver with a screw-holding feature (shown in the right-bottom corner of Fig. 8-16). When inserting the pins (as shown in Fig. 8-16), the search strategy described above is employed. Subsequently the robot grasps the screwdriver and brings it to the first screw. The screwdriver starts to rotate and engages the screw. The robot brings the screwdriver to the side hole (with chamfered corners) and pushes it in the hole direction. The combination of the rotary motion of the screwdriver and the linear motion of the robot in the X direction inserts the screw into the side hole. The force sensor synchronizes these two motions. The screw is pushed by the robot, but if the force exceeds a certain level, the robot motion is temporarily stopped. The complete insertion of the screw is detected by a sudden change in the screwdriver current, and this signals the robot to retract.

In addition to assembly tasks, force sensors are used in classification of parts by their weight, and in welding and deburring operations (see

FIG. 8-16 Assembly task performed with the aid of the Technion's force sensor.

Chap. 6). In the teaching stage of welding and in deburring, an unknown contour should be followed. In this case, one specifies the contact force which is to be maintained. Deviations from that force are used to control the position of the tool tip in the direction perpendicular to the contour—an increase in force causing motion away from the contour and vice versa. The force sensed parallel to the surface can be used to command rotations of the tool to maintain the desired orientation relative to the contour.

8.6 ARTIFICIAL INTELLIGENCE

Robots which operate effectively in a changing environment must be equipped with sensors and have some degree of artificial intelligence (AI). In order to discuss AI, it is first necessary to agree on what intelligence itself is, and that is not an easy step.

One of the definitions of *intelligence* is "the ability to respond quickly and successfully to a new situation."† Accordingly, a necessary condition to apply intelligence to robots is to make them capable of learning about the existence of a new situation in their working environment, and this must be accomplished by adding suitable sensors, such as vision and range detectors.

---

†With permission. From *Webster's New World Dictionary*, Second College Edition, Copyright © 1982 by Simon & Schuster, Inc.

In order for the robot to respond to a new situation, its computer must contain an AI algorithm. The subject of AI is a science by itself and is treated in many books [e.g., Winston (1977)]. In robotics, to prepare efficient AI algorithms the robot's designer must predict the various problems and situations that the robot might encounter in its environment and prepare a suitable algorithm to deal with each of these circumstances.

An example of an AI algorithm is the one used in the classification of parts, such as given in Fig. 8-5. The key to successful operation lies in establishing a suitable set of part matching rules. These rules and their relative weights in making the final decision regarding the part routing are the basis for the AI algorithm. The combination of sensors and AI algorithms determines the level of intelligence of the robot, which, in turn, provides its ability to respond successfully to new situations which occur in its environment. The smarter the robot's designer is, the higher the level of AI that the robot possesses.

Another definition of intelligence is "the ability to learn from experience and to acquire knowledge."† There are algorithms which enable self-learning of a robot. Let us assume a situation in which a robot which is equipped with a wrist force sensor has to insert pegs into holes in assembly tasks. The gripper is programmed to take a peg from a cartridge, move it a certain distance, and insert it into the hole. Because of the positional errors of the robot, the first peg will not initially reach the center of the hole, but with the aid of the force sensor and a search strategy, the robot eventually will find the center of the hole and will insert the peg. The incremental distances that the robot's joints had to move from the hole's programmed position to the final position can be stored in the robot's memory and used as correction values. When the robot picks up the next peg, the correction values are added to the programmed joint positions, and probably the peg will now reach the center of the hole at the outset. Even if the hole center is not initially reached, the correction distance is smaller than previously, meaning that the robot has learned from experience and acquired knowledge, or in other words, the robot possesses intelligence.

Although robots are capable of self-learning and of responding to new situations, they cannot possess the full range of human intelligence. I have frequently been asked: If robots are capable of self-learning, will they eventually take command of the world?

I agree that robots are about to cross a fuzzy line past which they will not only do things that they are programmed to but also things

†With permission. From *Webster's New World Dictionary*, Second College Edition. Copyright © 1982 by Simon & Schuster, Inc.

which require intelligence based on a set of rules which resides in the robot's computer. However, as smart as a robot can get, it can never be as intelligent as its progenitor.

Let us take for example the "common sense" which is frequently required when encountering new situations. We are so accustomed to using common sense that we take it for granted. But common sense demands knowing the exceptions to rules, and a machine which is programmed to judge new situations on the basis of programmed rules cannot have common sense. The fact that it takes natural intelligence to create the artificial kind gives humanity a perpetual edge, and the fear of being overtaken by robots seems to be unrealistic.

REFERENCES

Agin, G.: "Real Time Control of a Robot with a Hand-Held Camera," *9th Int. Symp. Expo. Ind. Robots*, Washington, D.C., March 1979, pp. 233–246.

Agin, G., and T. O. Binford: "Computer Description of Curved Objects," *Proc. 3rd Int. Joint Conf. Artif. Intell.*, Stanford, California, August 1973, pp. 629–640.

Barrow, H. G., and J. Tenenbaum: "Computational Vision," *Proc. IEEE*, vol. 69, no. 5, May 1981, pp. 572–595.

Binford, T. O.: "Sensor System for Manipulation," in E. Heer (ed.), *Proc. 1st Conf. Remot. Manned Syst. Explor. Oper. Space*, 1973, pp. 283–291.

Binford, T. O.: "Survey of Model-Based Image Analysis Systems," *Robot. Res.*, vol. 1, no. 1, Spring 1982, pp. 18–64.

Birk, J. et al.: "Image Feature Extraction Using Diameter-Limited Gradient Direction Histograms," *IEEE Trans. Pat. Anal. Mach. Intell.*, vol. PAMI-1, no. 2, April 1979, pp. 228–235.

Birk, J. et al.: *General Methods to Enable Robots with Vision to Acquire, Orient, and Transport Workpieces*, 6th Report, University of Rhode Island, August 1980.

Borenstein, J.: "Development of 3D Force Sensor," M.Sc. Thesis, Technion, Haifa, Israel, 1983.

Brady, M.: "Seeing Machines: Current Industrial Applications," *Mech. Eng.*, November 1981, pp. 52–59.

Duda, R. O., and P. E. Hart: *Pattern Classification and Scene Analysis*, Wiley-Interscience, New York, 1973.

Goto, T., K. Takeyasu, and T. Inoyama: "Control Algorithm for Precision Insert Operation Robots," *IEEE Trans. Syst. Man Cybern.*, vol. SMC-10, January 1980, pp. 19–25.

Kanade, T.: "Visual Sensing and Interpretation: The Image Understanding Point of View," *Comput. Mech. Eng.*, April 1983, pp. 59–69.

Karg, R., and O. E. Lanz: "Experimental Results with a Versatile Optoelectronic Sensor in Industrial Applications," *9th Int. Symp. Expo. Ind. Robots,* Washington, D.C., March 1979, pp. 247–264.

Kelley, R. et al.: "A Robot System which Feeds Workpieces Directly from Bins into Machines," *9th Int. Symp. Expo. Ind. Robots,* Washington, D.C., March 1979, pp. 339–356.

Levi, R.: "Multi-Component Calibration of Machine-Tool Dynamometers," *Trans. ASME, J. Eng. Ind.,* November 1972, pp. 1067–1072.

Nagel, R., C. Vanderburg, J. Albus, and E. Lowenfeld: "Experiments in Part Acquisition Using Robot Vision," *Robot. Today,* Winter 1980–81, pp. 30–40.

Nevins, J. L., and D. E. Whitney: "Computer Controlled Assembly," *Sci. Am.,* vol. 238, no. 2, 1978, pp. 62–73.

Nevins, J. L., and D. E. Whitney: "Assembly Research," *Ind. Robot,* March 1980, pp. 27–42.

Nitzan, D., A. E. Brain, and R. O. Duda: "The Measurement and Use of Registered Reflectance and Range Data in Scene Analysis," *Proc. IEEE,* vol. 65, February 1977, pp. 206–220.

Nitzan, D.: "Assessment of Robotic Sensors," *Proc. 1st Int. Conf. Robot Vision Sens. Cont.,* Stratford, U.K., April 1981, pp. 1–8.

Okada, T.: "Development of an Optical Distance Sensor for Robots," *Robot. Res.,* vol. 1, no. 4, Winter 1982–83, pp. 3–14.

Perkins, W. A.: "Area Segmentation of Images Using Edge Points," *IEEE Trans. Pat. Anal. Mach. Intell.,* vol. PAMI-2/No. 1, January 1980, pp. 8–15.

Rosen, C. A., and D. Nitzan: "Use of Sensors in Programmable Automation," *Computer,* December 1977, pp. 12–23.

Rosen, C. A., and G. J. Gleason: "Evaluating Vision System Performance," *Robot. Today,* Fall 1981, pp. 45–48.

Rosenfeld, A., and A. C. Kak: *Digital Picture Processing,* Academic Press, New York, 1976.

Salmon, M.: "Consideration on the Design of the Olivetti SIGMA: An Industrial Robot for the Manufacturing Industries," *6th Int. Symp. Ind. Robots,* March 1976, pp. 92–109.

Spalding, C. H.: "A 3-Axis Force Sensing System for Industrial Robots," *3rd Int. Conf. Assem. Autom.,* Stuttgart, W. Germany, May 1982, pp. 565–576.

Simunovic, S.: "Force Information on Assembly Processes," *5th Int. Symp. Ind. Robots,* Chicago, Ill., September 1975, pp. 415–431.

Thompson, A. M.: "Camera Geometry for Robot Vision," *Robot. Age,* March–April 1981, pp. 22–34.

Van Brussell, H., and J. Simons: "Automatic Assembly by Active Force Feedback Accommodation," *8th Int. Symp. Ind. Robots,* Stuttgart, W. Germany, 1978, pp. 181–193.

Vanderburg, C. J., J. S. Albus, and E. Barkmeyer: "A Vision System for Real

Time Control of Robots," *9th Int. Symp. Expo. Ind. Robots*, Washington, D.C., March 1979, pp. 213–232.

Van Voorhis, D. C., and T. H. Morzin: "Memory Systems for Image Processing," *IEEE Trans. Comput.*, vol. C-27, no. 2, February 1978, pp. 113–125.

Warnecke, H. J., M. Schweizer, and D. Haaf: "Programmable Assembly with Tactile Sensors and Visual Inspection," *1st Int. Conf. Assem. Autom.*, Brighton, UK, March 1980, pp. 23–32.

Watson, P. C., and S. H. Drake: "Pedestal and Wrist Force Sensors for Automatic Assembly," *Proc. 5th Int. Symp. Ind. Robots*, Chicago, Ill., September 1975, pp. 501–511.

Will, P. M., and D. D. Grossman: "An Experimental System for Computer Controlled Mechanical Assembly," *IEEE Trans. Comput.* vol. C-24, no. 9, 1975, pp. 879–888.

Winston, P. H.: *The Psychology of Computer Vision*, McGraw-Hill Book Co., New York, 1975.

Winston, P. H.: *Artificial Intelligence*, Addison-Wesley Publishing Co., Reading, Mass., 1977.

CHAPTER 9

Installing a Robot

Before a decision to purchase a robot is made, the prospective buyer should analyze the reasons for wanting it. What is the type of work to be done with the robot? Do you really need a robot or, possibly, hard automation equipment? Does your company have a long-range plan into which robotics fits? Have you considered the social impact in the plant?

The first thing to remember when acquiring a robot is that you look for a solution to a manufacturing problem. The robotized solution is considered most efficient when you can achieve one or more of the following benefits (see Sec. 1.2):

1. Reduce production cost because (a) robot work is cheaper than human labor, (b) robot programming flexibility can be used in varying batch sizes in manufacturing
2. Increase productivity
3. Improve product quality
4. Improve workers' lives by performing the undesirable jobs with robots

The first step in selecting a robot is to make a plant survey in order to find the best operations for automating with robots. Next, select the appropriate robots. Finally, an economic analysis should be carried out in order to find the payback on the investment in each of the candidate operations. If you are considering installing your first robot (Bergant, 1982), then the social impact might be the significant factor in your considerations rather than the shortest payback period. You might select an operation in which you remove people from hazardous and risky jobs, thereby avoiding possible negative reactions of workers and unions.

At the economic analysis stage, expenses for the robot working environment should be taken into account. You have to design a stand for the robot and a transfer system or feeders to bring materials to the robot and take away finished products. Sometimes design modifications are required in the product to enable assembly by a robot (see Sec. 6.5). In this case some robots require more product modifications than others, and this might be your major selection criterion.

Long-range company policy is another factor. The company must limit the number of robot types in the plant in order to facilitate maintenance problems. Therefore, providing the cheapest solution to one manufacturing problem does not mean that you have found the most cost-effective solution for the entire plant.

These are the considerations which are discussed in this chapter, with the purpose of assisting engineers in finding the appropriate robotic solution in their plant. The selection of robots for process applications [e.g., welding (Stauffer 1982a), injection molding (Stauffer, 1982b), and spray painting] is not discussed and considerations are mainly given for robots for handling, machine loading, forging, and assembly.

9.1 A PLANT SURVEY

The first step in the procedure of selecting a robot is to conduct a survey of the plant. During the initial plant walkthrough you will find many manufacturing operations which look like appropriate candidates for robot applications. If you want to make a preliminary check whether a robot can perform a specific job, you should (1) close your eyes, (2) use one hand only, and (3) imagine that you have a mitten on your hand. The mitten eliminates your sense of touch and emulates a jaw-type gripper. Now, if under these restrictions you think that you can perform the job, consider this application further. You will immediately find that the problem of orientation of the incoming parts is one of the major limits in applying robots. Once the part orientation problem has been solved, in many cases a piece of hard automation equipment will provide a better economic solution than a robot.

WORK STATION INFORMATION SHEET

Client:_____ Plant:_____ Date:___/___/198___

Operation:_____ Machine #:_____ Bay:_____

	ITEM			ITEM	
1	Number of different parts processed on machine		24	Can parts be loaded in same way as previous operation?	
2	Part name		25	Can parts be loaded in same way into next operation?	
3	Part number		26	Is operator judgement required?	
4	Part weight (range)		27	Is vision required?	
5	Part size (range)		28	What is line speed?	
6	Part material		29	What is line spacing?	
7	Part temperature		30	Are parts prints available (including tolerances)?	
8	Cycle time		31	Is work station layout available?	
9	Number of elements per cycle		32	Is fixture print (including tolerances available)?	
10	Load/unload time		33	What is ambient temperature range?	
11	Hourly production rate		34	Is environment dirty?	
12	What is the recorded efficiency?		35	Is environment corrosive?	
13	Number of shifts		36	Is environment dry/humid?	
14	Number of operators/shift		37	Is environment dangerous or unsuitable for human operator?	
15	Is the standard measured or negotiated?		38	Any sand, chip, liquid discharged in process?	
16	Are there any irregular operations?		39		
17	Are photographs of work stations available?		40		
18	Is operator using one or two hands?				
19	How is part presented (are drawings available)?			USE THIS AREA FOR ADDITIONAL NOTES	
20	How is part removed (are drawings available)?				
21	What is the required tolerance for handling?				
22	What is handling height above floor?				
23	What is the required part orientation?				

FIG. 9-1 A work station information sheet used in plant survey. (Quality Engineering.)

The initial plant walkthrough provides you with a coarse evaluation of all the potential robot applications in the plant. The next step is to conduct a comprehensive survey of these applications in order to collect the necessary data to apply robots. The data are accumulated on special forms, such as the one shown in Fig. 9-1. More comprehensive forms are available in the literature (Ottinger, 1981). These forms include information about various aspects: technical (items 4 to 9 in Fig. 9-1), economic (items 12 and 15,), environment (items 33 to 37), relationship to the adjacent operations in the line (items 24 and 25), and the line itself (28 and 29). There is information which is available from the supervisor (item 31), and from the operator performing the task (items 16 and 26). The operator can tell you about irregularities: how often parts cause assemblies to jam, how often parts are missing or incorrectly oriented, how often defective parts arrive, or how often a part sticks to the die in forging. This is valuable information that might be unnoticed as such problems occur infrequently. However, irregularities will cause robot failure.

Special consideration should be given to the cycle time of the operation, since at the final selection stage the speed of the robot must fit the measured cycle time.

Standard forms cannot be used for all applications because the required information varies according to the application being reviewed (assembly, handling, machine loading, etc.). You might have to develop other forms that better fit the operations to be evaluated.

It is recommended that the survey start where materials are received in the plant and follow the product flow through the plant. In this way you pay attention to the sequence of adjacent operations. Part orientation is one of the biggest problems in introducing robots into a plant and can sometimes be solved by observing adjacent manufacturing operations. If you want to robotize operations, it does not make sense to pile a batch of parts haphazardly in a tub or box at the end of a workstation and expect the robot at the next station to be able to pick up the parts one by one with the correct orientation. Therefore, if part orientation could be maintained at the unloading state of a prior operation, this would facilitate loading at the next workstation. This can be done by arranging the parts on pallets in boxes with specialized fixtures or bottoms, in vertical stacks, or transmitting them between guides. When designing a robotized station notice that whenever a robot is expected to pick up parts, they have to be presented to the gripper at the same pickup point (or at a known grid pattern) with the same orientation (at least in the present state of robot development).

When conducting the survey you have to remember that the objective is that the robot will do the job, and it does not necessarily mean that the robot must copy the exact sequence of current manual operations. You don't only have to replace labor, you want to increase productivity.

One good example is in die castings (Engelberger, 1980), where a common arrangement for quenching dies is to drop them into a tank of cold water and let a worker pick them up at a later stage. A robot, however, can not "fish" the die castings from the tank, where their position and orientation is not known. The solution in this case is to make the robot grasp the part, take it from the mold, dip it into the water tank without releasing it, take it out after a given interval, and load it into a trim press for the next operation.

There are cases in which the method of presenting parts to tools can be changed. It does not matter if the robot brings the part to the tool or carries the tool to the part. Sometimes by changing the location of a tool the installation of the robot is facilitated. Work can sometimes be arranged with one robot tending two machines in a work sequence. Many operations are done in the factory in a certain way just because the human operator always did it that way; change them if this will make an effective use of robots!

On the other hand, the replacement of a human operator might require additional devices to compensate for the robot's inability to see and feel. These devices could be feeders for arranging parts in a single orientation, or sensors which would prevent dangerous situations for the robot and people in its environment from occurring. Sensors can avoid, for example, the closing of a press before the robot arm is removed. Typical sensors are photoelectric devices, infrared detectors, pressure switches, etc.

9.2 SELECTING A ROBOT

After the plant survey is completed, the various applications have to be evaluated. Usually only a few robots will be introduced at a time. The plant management has to define the criteria for robot installations. Such criteria might be

1. Selection of an operation(s) which a robot will be able to perform on a continuous basis (i.e., low occurrence of irregular elements) without major changes in the process

2. Ease of introduction to the plant from maintenance and engineering point of view (e.g., limited number of grippers and extra feeders)

3. Long-range planning of standard equipment in the plant (e.g., all robots should be dc-motor–driven if the payload is more than 3 kg)

4. Availability of enough work to ensure two-shift operation year-round, taking into account the increased productivity with robots

5. Return on investment should meet a certain limit of payback period or internal rate of return (see Sec. 9.3).

The next stage is to find a commercial-type robot most suitable for the selected operation(s) (Ottinger, 1982). There are process applications (e.g., spray painting and arc welding) which require special types of robots, and in these cases the selection process is relatively easy. In other applications, such as handling, machine loading, and assembly, the robot characteristics must fit the application requirement. The main problem, however, is that a robot system has many characteristics, and without using a systematic approach the selection procedure becomes difficult.

One common approach is to start with preparing a table containing the various robot models available in the market and their main characteristics. The second step is to decide upon a dominant characteristic to start the selection process with, and then to build a decision tree based on other characteristics. In many handling and loading applications the payload is the dominant feature, followed by the robot reach envelope. In assembly, the speed of operation is frequently chosen as the most important feature, and the robot selection process starts by choosing a group of robots having a certain permissible speed limit.

The next step is to find whether a certain robotic coordinate system is excluded because it does not fit the required application. For example, a cartesian coordinate robot cannot be used in most machine-loading and handling jobs. Together with the coordinate system, the minimum number of axes of motion and repeatability are also an important criterion. Next, environment considerations are taken into account (e.g, ambient temperature) and so on. At each of these decision steps the number of candidate robots is narrowed down, and at the end of the process only a few robots remain. Special factors which cannot be tabulated are then considered, and finally an economic analysis is carried out in order to make the final decision about the most appropriate robot.

Table 9-1 serves as an example of gathering commercial robot characteristics in a tabulated format. It contains more than 100 models of robots produced by 33 manufacturers which were available in the American market in 1982 (See Appendix B for full names and addresses of manufacturers.) The first part of the table contains the following information.

Control: CP   =   continuous path (Sec. 2.1.2)

PTP   =   point to point (Sec. 2.1.1)

PTPL   =   PTP, but with all axes simultaneously reaching the end point; if the distances between end points are small, the obtained trajectory is close to a straight line.

Drives: EL   =   electric servomotors (Sec. 3.2)

HYD   =   hydraulic actuators (Sec. 3.1)

PNM   =   pneumatic actuators

Programming Method: MAN = manual teaching (Sec. 7.1)

LT = lead-through teaching (Sec. 7.2)

L(AML) = programming language is AML (Sec. 7.3)

L(VAL) = programming language is VAL (Sec. 7.3)

Coordinate System: ART = articulated (Sec. 2.3.4)

CRT = cartesian (Sec. 2.3.1.)

CYL = cylindrical (Sec. 2.3.2)

SPH = spherical (Sec. 2.3.3)

Axes: number of available axes

Repeatability: the end of arm repeatability given in millimeters.

Load: maximum payload of the arm in kilograms.

Weight: The weight of the manipulator in kilograms.

Applications:

AD	= adhesive		G	= grinding
AS	= assembly		I	= inspection
AW	= arc welding		LD	= loading/unloading
CO	= coating			and handling
DB	= deburring		PN	= spray painting
DR	= drilling		SW	= spot welding

The second part of the table contains information about the manipulator: the arm and the wrist range and speed. The range information can assist in calculating the work envelope required to establish the robot position.

The first column indicates the horizontal range of the arm, and the next six columns the range and speed of the individual axes. For cartesian coordinate robots these six columns show the maximum ranges of the X, Y, and Z axes, and the maximum speed at each axis. For cylindrical coordinate robots, the first two of these six columns $(X;\theta_1)$ provides the base rotation in degrees, and the maximum corresponding speed in degrees per second; the next two columns $(Y;\theta_2)$ show the vertical range in millimeters and the maximum vertical speed; the two col-

TABLE 9-1 Characteristics of Commercial Robots (1982)

Manufac-turer	Model	Control	Drives	Program	Coor-dinates	Axes	Repeat-ability, mm	Load, kg	Weight, kg	Applica-tion
Adv. Robot	CYRO 750	PTPL	ELC	MAN	CRT	5	0.2		2613	AW
Adv. Robot	CYRO2000	PTPL	ELC	MAN	CRT	5	0.4		5500	AW
Adv. Robot	CRYO 820	CP	ELC	MAN	ART	5	0.2	10	300	AW
Amer. Rob.	MERLIN	CP	STP	LT MAN	ART	6	0.025	9	24	LD AS I W
ASEA	IRB-6	PTP	ELC	MAN	ART	5	0.2	6	125	LD AW
ASEA	IRB-60	PTP	ELC	MAN	ART	5	0.4	60	750	LD SW
ASEA	A30-A	PTPL	ELC	MAN	ART	5	0.2	6		DB AW AD
ASEA	IRB-60S	PTPL	ELC	MAN	ART	6	0.4	60		SW DB
ASEA	MHU SEN	PTP	PNM	MAN	CYL	3	0.1	15		MO LD
ASEA	MHU JUN	PTP	PNM	MAN	CYL	3	0.1	5		LD
Automatix	FANUC	PTPL	ELC	MAN	ART	5	0.2	10	230	AW DB FN
Bendix	ROBOT	PTP	ELC	MAN	ART	6	0.08	13.6/68		LD
Bendix	ML-360	PTP	ELC	MAN	ART	6	0.13	68		LD AS W I
Bendix	AS-160	PTP	ELC	MAN	SPH	6		20.5		AS LD
Binks	88-800	CP	HYD	LT	CYL	6	3.2	8.2		PN
Cin. Milac.	T3-566	PTPL	HYD	MAN	ART	6	1.25	45	2270	LD SW AW
Cin. Milac.	T3-726	PTPL	ELC	MAN	ART	6	0.15	20	436	LD AW DR
Cin. Milac.	T3-746	PTPL	ELC	MAN	ART	6	0.25	20	1800	LD AW DR
Cin. Milac.	HT3	PTP	HYD	MAN	ART	6	1.25	102		LD SW AW
Copperweld	CR-10	PTP	PNM	MAN	CYL	4	0.08	4.5		LD
Copperweld	CR-50	PTP	PNM	MAN	CYL	4	0.25	11.3		LD
Copperweld	CR-5	PTP	PNM	MAN	CYL	4		2.5		LD
Copperweld	CR-100	PTP	PNM/ELC	LT	CRT	3	0.05	5.5		LD AS
Cybotech	H80	CP	HYD	MAN	CYL°	6	0.2	80	1850	LD SW AW
Cybotech	V80	CP	HYD	MAN	ART	6	0.2	80	1800	LD SW AW
Cybotech	G80	CP	HYD	MAN	CRT	6	0.2	80	4100	LD SW AW
Cybotech	P15	CP	HYD	MAN	ART	7	5	15	750	PN CO
Cybotech	V30	CP	HYD	MAN	ART	6	0.2	30	600	LD SW AW
Cybotech	H8	PTP	ELC	MAN	CYL	5	0.1	8		AS LD I
DeVilbiss	TR-3500	CP/PTP	HYD	LT	ART	6	1.0			PN CO AD
DeVilbiss	TR-3500W	CP/PTP	HYD	LT	ART	5	0.5			AW AS DR DB AD
GCA	XR	CP	ELC	L(MDI)	CRT	6	0.5	1130		DB LD AS
GCA	B1440	CP	ELC	L(MDI) LT	CYL	6	0.5	50	900	LD AW SW
GCA	B2600	CP	ELC	L(MDI) LT	CYL	6	1.0	100	1200	LD AW SW
GCA	B4700	CP	ELC	L(MDI) LT	CYL	6	1.0	350	2500	LD AW SW
GCA	P800	CP	ELC	L(MDI)	ART	5	0.5	30	180	AS LD SW

TABLE 9-1 Characteristics of Commercial Robots (1982) *(Continued)*

Manufac-turer	Model	Range, mm	Range, $X;\theta_1$	Speed, $X;\theta_1$	Range, $Y;\theta_2$	Speed, $Y;\theta_2$	Range, $Z;\theta_3$	Speed, $Z;\theta_3$	Roll	Pitch	Yaw
					Arm					Wrist	
Adv. Robot	CYRO 750	2032	2032 mm	254 mm/s	762 mm	169 mm/s	762 mm	169 mm/s		130	720
Adv. Robot	CYRO2000	2032	2032 mm	127 mm/s	2032 mm	127 mm/s	2032 mm	63 mm/s		130, 90	720, 90
Adv. Robot	CYRO 820	1305	240°	90°/s	820 mm	780 mm/s	890 mm	1000 mm/s	380	200	
Amer. Rob.	MERLIN	1000	360°		360°		300°		Con-tinuous	180	Con-tinuous
ASEA	IRB-6	1159	340°	95°/s	80°	0.75 m/s	65°	1.1 m/s	360, 195	180, 115	
ASEA	IRB-60	2288	330°	90°/s	70°	1.0 m/s	65°	1.35 m/s	360, 150	195, 90	
ASEA	A30-A										
ASEA	IRB-605										
ASEA	MHU SEN	1100	Con-tinuous	90°/s	1100 mm	1 m/s	500 mm	0.3 m/s			
ASEA	MHU JUN	500	200°	180°/s	500 mm	1 m/s	150 mm	0.5 m/s			
Automatix	FANUC	1257	300°		95°		70°		370, 180	175, 120	
Bendix	ROBOT										
Bendix	ML-360	2540	300°	90°/s	98 in	56 in/s	165 in	52 in/s	, 155	, 103	, 103
Bendix	AS-160										
Binks	88-800	1220	130°	60°/s	48 in	30 in/s	84 in	30 in/s	270, 150	180	180
Cinc. Milac.	T3-566	2464	240°		2464 mm		3911 mm		240	180	180
Cinc. Milac.	T3-726		285°		1040 mm		1600 mm		Con-tinuous	240	240
Cinc. Milac.	T3-746		270°		2464 mm		3300 mm		Con-tinuous	240	240
Cin. Milac.	HT3	2591	240°		2591 mm		3911 mm		270	180	180
Copperweld	CR-10		304 mm	413 mm/s	200°	190°/s	51 mm	5 mm/s	180		
Copperweld	CR-50		457 mm	586 mm/s	200°	32°/s	127 mm	7 mm/s	180		
Copperweld	CR-5										
Copperweld	CR-100	609	609 mm	203 mm/s	609 mm	203 mm/s	305 mm	203 mm/s			
Cybotech	H8O	2200	270°	57°/s	270°	57°/s	1600 mm	660 mm/s	335, 172	210, 172	344, 172
Cybotech	V80	2200	270°	57°/s	210°	57°/s	270°	57°/s	335, 172	210, 172	344, 172
Cybotech	G80	3000	3000 mm	1 m/s	1500 mm	1 m/s	1000 mm	0.5 m/s	335; 172	210, 172	344, 172
Cybotech	P15	2700	230°		80°		90°		340	180	180
Cybotech	V30	2000	240°	57°/s	80°	57 °/s	90°	57°/s	380, 57	240, 57	280, 57
Cybotech	H8										
DeVilbiss	TR-3100	8010	93°		75°		72°		210	210	210
DeVilbiss	TR-3500W	8010	93°		75°		72°		210	210	210
GCA	XRTM	12192	9100 mm	915 mm/s	3000 mm	915 mm/s	2100 mm	57°/s	33, 57	210, 57	330, 57
GCA	B1440	1780	300°	60 °/s	1000 mm	600 mm/s	700 mm	600 mm/s	180, 60	120, 60	270, 60
GCA	B2600	2570	270°	60 °/s	1480 mm	600 mm/s	1350 mm	600 mm/s	180, 180	120, 60	270, 60
GCA	B4700	2570	270°	40 °/s	1480 mm	400 mm/s	1350 mm	400 mm/s	180, 45	120, 45	270, 45
GCA	P800	1360	270°	60 °/s	180°	60°/s	300°	66°/s	200, 75	200, 75	

TABLE 9-1 Characteristics of Commercial Robots (1982) *(Continued)*

Manufac-turer	Model	Control	Drives	Program	Coordinate	Axes	Repeat-ability, mm	Load, kg	Weight kg	Applica-tion
GCA	P300V	CP	ELC	L(MDI)	ART	5	0.1	5	90	AS LD SW
GCA	P500	CP	ELC	L(MDI)	CYL	5	0.2	7	150	AS LD SW
GCA	F —	CP	ELC	L(MDI)	ART/ CRT	5	0.5	10	350	AW SW
G.E.	ALLEGRO	PTP	ELC	L(HELP)	CRT	6	0.025	6.5		AS LD
G.E.	GP 66	PTP	ELC	MAN	SPH	6	1.0	30	875	LD SW
G.E.	GP 132	PTP	ELC	MAN	ART	7	0.75	60	1000	LD SW
G.E.	S-6	PTPL	ELC	MAN	CYL	6	5.0	3	500	PN
G.E.	AW7	CP	HYD/ ELC	MAN	CRT	5	1.0		1300	AW
G.E.	MH 33	PTP	ELC	MAN	SPH	4	2.5	15	730	LD
G.E.	P5	PTPL	ELC	MAN	ART/ CRT	5	0.2	10	230	AW DB
G.E.	P6	CP	ELC	MAN	ART	6	1.0	15	760	AW DE G
G.E.	SW 220	PTP	ELC	MAN		5	2.5	100	3800	SW LD
Gen. Numeric	GN0	PTP	ELC	MAN	CYL	6	0.5	20	100	LD
Gen. Numeric	GN1	PTP	ELC	MAN	CYL	5	1.0	20	650	LD
Gen. Numeric	GN3	PTP	ELC	MAN	CYL	5	1.0	78	1700	LD
Graco	OM 5000	CP	HYD	MAN						PN
Hobart	L10	CP	ELC	MAN	ART	5	0.2	10	400	AW
IBM	RS-1	CP	HYD	L (AML)	CRT	6	0.2	2.3	910	ASI
IBM	7535	PTP	ELC	L(AML)	ART	4	0.05	6	60	AS LD I
Ind. Automates	AUTO-MATE	PTP	PNM/ HYD	MAN	CYL	4	0.38	4.5		LD
IRI	IRI	PTP	PNM	LT MAN	ART	5	1	23	91	LD
Kuka	IR 60	PTP	ELC	L MAN	ART	6	1	100	2000	LD SW AW
Kuka	IR 200	PTP	ELC	L MAN	ART + CRT	7	1	60	2500	LD SW
Mobot	MOBOTS	{ PTP	ELC OR HYD	MAN	CYL OR CRT					LD }
Nachi	UM8600AK	PTPL	ELC	MAN	ART	6	1	50	980	SW
Pickomatic	FR 100-5	PTP	ELC	MAN	ART	5	0.6	30	325	
Pickomatic	FR 200-3	PTP	ELC	MAN	CYL	3	0.4	70	200	
Pickomatic	FR 300-3	PTP	ELC	MAN	CYL	3	0.4	30	300	
Planet	ARMAX	PTP	ELC	MAN	CYL	2–7	0.5	68		LD DR
			ELC		SPH	3			200	
Positech	PROBOT	{ PTP	HYD	MAN	OR	to	0.5	to	to	LD PN }
			PNM		CYL	7		570	1500	
Prab	4200	PTP	HYD	LT	SPH	6	0.2	34		LD
Prab	5800 HD	PTP	HYD	LT	SPH	6	0.2	45		LD
Prab	MODEL E	PTP	HYD	LT	CYL	5	0.76	45		LD

TABLE 9-1 Characteristics of Commercial Robots (1982) *(Continued)*

Manufac-turer	Model	Range, mm	Range, $X;\theta_1$	Speed, $X;\theta_1$	Range, $Y;\theta_2$	Speed, $Y;\theta_2$	Range, $Z;\theta_3$	Speed, $Z;\theta_3$	Roll	Pitch	Yaw
GCA	P300V	711	270°		180°	90°/s	300°	90°	270, 90	210, 90	
GCA	P500	1194	200°	75 °/s	180°	60°/s	26 in	24 in/s	270, 75	210, 75	
GCA	F	1585	1560 mm	600 mm/s	1470 mm	600 mm/s	90°	45°/s		6 in (Z)	120, 60
G.E.	ALLEGRO	1300	1300 mm	670 mm/s	310 mm	670 mm/s	265 mm	670 mm/s	360, 132	196, 132	360, 132
G.E.	GP-66	2000	320°	80 °/s	65°	30°/s	700 mm	450 mm/s	350, 120	350, 180	350, 180
G.E.	GP-132	2405	320°	59 °/s	140°	50°/s	120°	87°/s	345, 144	210, 142	540; 140
G.E.	S6	1950	150°		1320 mm	1 m/s	3111 mm	1 m/s	250	250	250
G.E.	AW7	1500	1500 mm	16.5 mm/s	1000 mm	16.5 mm/s	1100 mm	16.5 mm/s		55	200
G.E.	MH 33	2500	320°	160 °/s	20°	30°/s	1800 mm	1.2 m/s			
G.E.	P5	1257	300°		95°		70°		370, 180	175, 120	
G.E.	P6	2200	320°	80 °/s	65°	30°/s	100°	50°/s	350, 120	270, 120	
G.E.	SW 220	3000	2000 mm	250 mm/s	40°	6°/s	1000 mm	250 mm/s	340, 90	80, 80	
Gen. Numeric	GNO	150	120°	120 °/s	150 mm	508 mm/s	150 mm	508 mm/s	180, 90	180, 120	120, 120
Gen. Numeric	GN1	1100	300°	60 °/s	1100 mm	1.02 m/s	500 mm	508 mm/s	270, 60	10, 30	
Gen. Numeric	GN3	1200	300°	60 °/s	1200 mm	1.02 m/s	1200 mm	508 mm/s	300, 80	190, 80	
Graco	OM 5000										
Hobart	L10	1186	240°	90 °/s	80°	800 mm/s	60°	1.1 m/s	360, 150	180, 100	
IBM	RS-1	1473	457 mm	1 m/s	1473 mm	1 m/s	431 mm	1 m/s	270, 180	180, 180	270, 180
IBM	7535	650	200°	1.45 m/s	160°	1 m/s	3 in		360		
Ind. Automates	AUTO-MATE										
IRI	IRI	2000	Con-tinuous	60°/s	180°	60°/s	230°	60°/s	Cont. 120	240, 120	
Kuka	IR 60	2750	320°	76°/s	1500 mm	1.2 m/s	2500 mm	2 m/s	360, 120	360, 120	360, 120
Kuka	IR 200		6000 mm	1.2 m/s	320°	160°/s	180°	125°/s	180, 125	350, 200	180, 200
Mobot	MOBOTS	Modular, unlimited range									
Nachi	UM8600AK	1351	270°	50°/s	90°	50°/s	65°	50°/s	360, 90	360, 90	360, 90
Pickomatic	FR 100-5	1370	340°	90°/s	230°	60°/s	270°	135°/s	360,	Cont. 150	
Pickomatic	FR 200-3	1500	340°	90°/s	800 mm	1 m/s	500 mm	500 mm/s			
Pickomatic	FR 300-3	710	500 mm	500 mm/s	40°	120°/s	300°	120°/s	300, 120	360, 360	
Planet	ARMAX	2590	270°		54 in		10 in				
Positech	PROBOT		360°	60°/s			2540 mm	460 mm/s	360	180	300
Prab	4200	2083	300°		1067 mm		810 mm		270	180	180
Prab	5800 HD	2489	300°		1473 mm		1000 mm		180	180	180
Prab	MODEL E	1343	240°		1067 mm		762 mm		270	180	240

TABLE 9-1 Characteristics of Commercial Robots (1982) *(Continued)*

Manufac-turer	Model	Control	Drives	Program	Coor-dinates	Axes	Repeat-ability, mm	Load, kg	Weight, kg	Applica-tion
Prab	MODEL FA	PTP	HYD	LT	CYL	7	1.27	110		LD SW
Prab	MODEL FB	PTP	HYD	LT	CYL	7	1.27	270		LD SW
Prab	MODEL FC	PTP	HYD	LT	CYL	7	2.0	910		LD SW
Reis	RR625	CP	ELC	MAN LT L	ART	6	0.20	55	1875	LD SW
RHINO	XR-1	PTP	ELC	MAN	ART	5	1.35	0.5	6	Education
Seiko	700	PTP	PNM	SWITCHES	CYL	4	0.64	1.0		LD AS
Seiko	200	PTP	PNM	SWITCHES	CYL	3	0.01	0.7		AS LD
Seiko	400	PTP	PNM	SWITCHES	CRT	3	0.025	4		LD
Seiko	100	PTP	PNM	SWITCHES	CRT	3	0.01	1.3		AS LD
Thermwood	CA5 510P				CRT	5+				AS DB DR
Thermwood	SERIES 3	PTP	HYD	MAN	ART	5+	1.5	23		LD
Thermwood	SERIES 6	CP	HYD	LT	ART	6+	3.2	8		PN CO
Thermwood	SERIES 7	CP	HYD	LT	ART	6+	1.5	11		LD
Toshiba	TOS.IX12	PTP	HYD	MAN L	SPH	5	1.0	10		LD SW
Toshiba	TOS.IX15	PTP	HYD	MAN L	SPH	6	1.0	20		LD SW
Unimation	1000	PTP	HYD	MAN	SPH	5	1.27	23		LD
Unimation	2000B	CP	HYD	MAN	SPH	6	1.27	134		LD SW AW
Unimation	2100B	CP	HYD	MAN	SPH	6	2.0	134		LD SW AW
Unimation	2000C	CP	HYD	MAN	SPH	6	1.27	134		LD SW AW
Unimation	2100C	PTP	HYD	MAN	SPH	6	2.0	123		
Unimation	4000B	PTP	HYD	MAN	SPH	6	2.0	205		SW
Unimation	APPRENT.		ELC			5	1.0		34	AW
Unimation	PUMA 250	CP	ELC	L(VAL)MAN	ART	5	0.05	1.5	7	LD AS
Unimation	PUMA 600	CP	ELC	L(VAL)MAN	ART	6	0.10	2.5	55	LD AS
United Tech.	NIKO 25	PTP	ELC	MAN	ART	5	0.25	2.5	130	
United Tech.	NIKO 50	PTP	ELC	MAN	ART	5	0.25	5	240	
United Tech.	NIKO 200	PTP	ELC	MAN	CRT	6	0.35	20	1400	SW
United Tech.	NIKO 600	PTP	ELC	MAN	CRT	6	0.35	60	1650	SW
Westinghouse	SER. 5000	PTP	ELC	LT L	CRT	9	0.10	10	1500	AS I

TABLE 9-1 Characteristics of Commercial Robots (1982) (*Continued*)

Manufacturer	Model	Arm							Wrist		
		Range, mm	Range, $X;\theta_1$	Speed, $X;\theta_1$	Range, $Y;\theta_2$	Speed, $Y;\theta_2$	Range, $Z;\theta_3$	Speed, $Z;\theta_3$	Roll	Pitch	Yaw
Prab	MODEL FA	1994	270°		1524 mm		1524 mm		270	180	240
Prab	MODEL FB	2100	270°		1524 mm		1524 mm		270	180	240
Prab	MODEL FC	2273	270°		1524 mm		1524 mm		270	180	240
Reis	RR625	3400	270°	90°/s	360°	90°/s	1200 mm		Continuous		270
RHINO	XR-1	572	360°	27°/s	180°	23°/s	270°	23°/s	360,	360,	
Seiko	700	200	120°		200 mm		40 mm		180		
Seiko	200	370	90°				10 mm		180		
Seiko	400	370	400 mm				80 mm		180		
Seiko	100	410	200 mm				50 mm		180		
Thermwood	CA5 510P	3050	120 in	50 in/s	60 in	50 in/s	24 in	50 in/s	360, 36	240, 36	
Thermwood	SERIES 3	1524	280°	90°/s	100°	90°/s	280°	90°/s	360	210	
Thermwood	SERIES 6	3050	135°	60°/s	48 in	30 in/s	84 in	30 in/s	270, 150	180	
Thermwood	SERIES 7	1550	280°	45°/s	39 in	30 in/s	76 in	30 in/s	300, 150	180	300, 150
Toshiba	TOS.IX12	1900	220°	90°/s	60°	30°/s	700 mm	700 mm/s	220, 90	220, 90	
Toshiba	TOS.IX15	2515	220°	90°/s	60°	30°/s	900 mm	700 mm/s	360, 90	220, 90	360, 90
Unimation	1000	2290	208°		56°		42 in			90	90
Unimation	2000B		208°		56°		42 in		360	211	200
Unimation	2100B		208°		46°		54 in		360	211	200
Unimation	2000C		208°		56°		42 in		360	220	200
Unimation	2100C		208°		46°		54 in		360	220	200
Unimation	4000B		200°		51°		52 in		370	226	320
Unimation	APPRENT.	1143	90°		50°		35 in			175	180
Unimation	PUMA 250	406	315°	109°/s	310°	103°/s	300°	150°/s	575, 500	240, 320	525, 300
Unimation	PUMA 600	914	320°	92°/s	250°	57°/s	270°	137°/s	300, 258	200, 258	532, 270
United Tech.	NIKO 25		320°	90°/s	240°	90°/s	250°	110°/s	210, 110	350, 280	
United Tech.	NIKO 50		320°	90°/s	240°	90°/s	290°	110°/s	210, 110	350, 180	
United Tech.	NIKO 200	5500	6000 mm	0.75 m/s	2000 mm	0.75 m/s	1000 mm	0.30 m/s	340, 120	340, 120	340, 120
United Tech.	NIKO 600	5500	6000 mm	1 m/s	2000 mm	1 m/s	1200 mm	0.50 m/s	340, 120	340, 120	340, 120
Westinghouse	SER. 000	1473	1473 mm	533 m/s	406 mm	533 mm/s	406 mm	533 mm/s	360,	90,	

umns $(Z;\theta_3)$ provide the horizontal range and speed. For spherical robots, the first and second columns (of the six) give the base rotation angle and speed; the third and fourth, the elevation angle and speed; the fifth and sixth, the reach and speed of the telescoping linear axis. For articulated robots the six columns give the base, shoulder, and elbow angular ranges and speeds.

The last three columns give the wrist angular rotation in degrees, and, if available, the corresponding maximum speed in degrees per second. For example, the number 180, 115 means a rotation range of 180° at maximum speed of 115°/s.

The variables given in the table are sometimes dependent. For example, in some robots the maximum payload can be accommodated only when the robot is moving at a reduced speed and with the arm not completely extended. Therefore, always allow a safety margin for specifications provided by the robot manufacturer.

In machine-loading applications (not machine tools) there is sometimes a trade-off between cycle time and maximum payload. This happens when several parts (typically two to four) are loaded simultaneously; the engineer has to select between loading one part at a time (small payload and large loading time); loading two parts at a time by a double gripper (a double payload but almost half the loading time), or loading more parts by a multijaw gripper. The robot motion speed will probably be slightly reduced when increasing the payload.

Using the procedure described above, several robots from different vendors will probably be encountered as candidates for each application, and the final selection is based upon an economic analysis. Before proceeding to this stage a decision is made as to the type of gripper and the required peripheral equipment such as a riser, feeders, etc. Their prices are part of the robot installation cost and must be included in the economic analysis.

9.3 ECONOMIC ANALYSIS†

In order to analyze the economic justification of installing a robot, we have to consider the total *cost* and the anticipated *savings* of the robot system.

When considering the robot cost you have to make sure to include the total cost of the robot system. In addition to the direct robot cost, there are other costs, such as:

†This section was contributed by A. Varkovitzky, President of Quality Engineering Company, Southfield, Michigan.

1. Robot gripper mechanism, which is usually priced separately
2. Sensors, both in the robot and in the machine (to ensure parts are presented in the right location and orientation and that all tools are in the correct position)
3. Layout changes necessary for installation of the robot and supply of the necessary utilities (electricity, compressed air, etc.)
4. Feeders or other devices necessary to present parts in a predetermined orientation
5. Maintenance cost (a good rule of thumb for estimating annual maintenance cost is 10 percent of the robot cost).

Savings could be generated in several ways:

1. *Labor Replacement.*   The cost should include direct pay as well as fringe benefits. A robot could replace one or more operators in one, two, or three shifts.
2. *Quality Improvement.*   A robot's work is more reliable and consistent than that of a human operator. This could lead to an improvement in quality in some applications. Wherever possible, it is a good idea to try to quantify this improvement by multiplying the number of additional acceptable parts anticipated in 1 year by the part cost ($Q = N \times c$).
3. *Increase in Productivity.*   In operations where cycle time was determined by the speed of the human operator rather than machine time, a robot may reduce cycle time and increase productivity. Although there were estimates of 25 percent increase in productivity by using robots, it is difficult in general to quantify the value of the increase. It is, however, especially important to consider it in cases where the productivity increase can save capital investment in additional machinery.
4. *Indirect Savings.*   A robot may produce indirect savings, such as savings in materials, reduced scrap or rework, and reduced workers' compensation costs because of injuries.

The cost and saving items discussed above are

COST:

R	= Robot cost	D	= Layout changes
G	= Robot gripper	F	= Feeders
S	= Sensors	M	= Annual maintenance cost

Annual Savings:

$$L = \text{Labor saving}$$

$$Q = \text{Quality improvement}$$

$$I = \text{Production increase}$$

This is by no means an inclusive list and individual attention should be paid to every application.

The total capital investment C is

$$C = R + G + S + D + F \qquad (9\text{-}1)$$

The total annual savings A are

$$A = L + Q + I \qquad (9\text{-}2)$$

There are two common methods used by industry to analyze economic justification of robots: the payback period and the return on investment.

9.3.1 The Payback Period

In this method, the plant management wants to know the length of time required for paying back the initial investment. Every company will have a rule of thumb as to what payback period is acceptable. The payback period P (in years) is the total investment divided by the net annual savings.

$$P = \frac{C}{A - M} \qquad (9\text{-}3)$$

EXAMPLE 9-1.

A robot used for machine loading is priced at $46,000, the special gripper required for it costs $5000, the sensors cost $1000, and the feeder cost is $30,000. There are no layout changes. The robot will replace one operator. The operator's rate is $16 per hour including fringe benefits and the operator works 250 days a year, 8 hours a day. No production increase or quality improvements are anticipated. What is the payback period for one-shift and for two-shift operation?

Solution. The values of the various variables are

$$R = \$46,000 \qquad L = \$16/\text{hour} \times 250 \text{ days/year} \times 8 \text{ h/day}$$

$$G = \$5000 \qquad\qquad = \$32,000/\text{year/shift}$$

$$S = \$1000 \qquad Q = 0$$

$$D = 0 \qquad I = 0$$

$$F = \$30,000 \qquad M = \$4600 \ (10\% \text{ of } \$46,000)$$

According to Eq. (9-1), the total cost is

$$C = R + G + S + D + F = \$82,000$$

According to Eq. (9-2) the total annual savings are

$$A = L + Q + I = \$32,000 \text{ for one shift, and } \$64,000 \text{ for two-}$$

shift operation

The payback period is calculated by Eq. (9-3) for one-shift operation

$$P = \frac{82,000}{32,000 - 4,600} = 2.99 \text{ years}$$

and for two-shift operation:

$$P = \frac{82,000}{64,000 - 4,600} = 1.38 \text{ years}$$

This result, a payback period of less than 2 years, is typical for most robots.

9.3.2 Return on Investment

Return on investment (ROI) is a more accurate approach and allows the management to compare anticipated return with other investment alternatives, both inside and outside the corporation. Also, if money for the capital investment has to be borrowed, it is easy to compare the anticipated return with the cost of the money. Two new variables are introduced:

n = the number of years used as the planning horizon ($n = 5$ is frequently used because of the practical difficulties of projecting beyond this limit, and because 5 years is used in the United States for depreciation purposes).

i = interest rate, and also the ROI.

The relationship between i, n, C, A, and M can be shown in the following formula:

$$A - M = C \frac{i(1 + i)^n}{(1 + i)^n - 1} \tag{9-4}$$

Equation (9-4) must be solved for i. This could be achieved by a trial-and-error method, but it is usually solved by a computer program or special-purpose calculators. Many programs have been written that also include the effects of residual value, investment tax credit, depreciation, and after-tax savings.

The solution of Example 9-1 for a 5-year planning horizon and no residual value yields an ROI as follows:

$$ROI = 20\% \text{ for one-shift operation}$$

$$ROI = 67\% \text{ for two-shift operation}$$

Notice, however, that the introduction of the first robot into a plant should not always be done on an economic basis alone. Considerations such as ease of installation, high confidence of success, introducing robot to employees, and education of maintenance people should play an important role in the decision process. With the first installation you are not only looking for success, but you are laying the groundwork for additional robot usage. A project failure resulting from haste in selecting the operation or the robot is not only a waste of capital, but usually postpones any further consideration of robots for several years.

9.4 A CASE STUDY

A high-volume automotive parts manufacturer decided to introduce robots into the plant. The plant is modern with different degrees of automation levels utilized to increase productivity. However, at the time of the study, robots were not operating in the plant. The objective of the study was to help the manufacturer to introduce the first robot into the plant. This included selecting the most suitable operation for the first robot and choosing a commercially available robot for this application.

Based upon preliminary discussions with the plant management personnel, several possible manufacturing operations for robotizing were suggested, and the following criteria were set:

1. Sufficient volume, with production on a two-shift basis each day
2. Ease of introduction to the plant from acceptance, maintenance, and engineering points of view
3. The payback period must be less than 2 years based upon two-shift operation
4. The selection of the robot will be based on its performance as well as on the availability of other models of robots from the same vendor (For the purpose of future standardization of robots in the plant)
5. An adequate amount of space around the selected machine for the robot without major rearrangements
6. Safety as a primary goal

The study included an initial walk through the plant to search for more possible operations to robotize, followed by a thorough survey.

Altogether about a dozen operations were found as good candidates for introducing the first robot into the plant, and many others which could be robotized in the far future. These operations were inspected by a project team, and a workstation information sheet (Fig. 9-1) was filled out for each of them. Drawings of parts, fixtures, machine layouts, and a plant layout were obtained for all of these operations, and a discussion was held on each application.

The operation which was found most suitable for the first robot installation was a machine-loading task, in which an operator picks up four parts at a time and loads them onto the machine by a twist-and-press motion. The parts are brought in a tub, which is set next to the operator. During the year, the machine handles three types of parts. However, the shape and weight of these parts vary only slightly with the weight of the heaviest part being about 1.1 lb (0.5 kg).

The reasons for selecting this operation were

1. Low occurrence of irregular parts
2. All parts could be handled with one feeder and one gripper
3. Volume of parts is high enough to keep operation going two shifts per day, year-round
4. The payback period was found to be 1.2 years. (See analysis at the end of this section.)
5. Productivity increase of 30 percent (The cost-benefit ratio associated with it was not included in the economic analysis, since the shorter machine time does not contribute at present to the productivity of the entire line. It has, however, future benefits.)

The next step was the selection of the robot for the loading operation. The initial criteria were a noncartesian coordinate robot with at least four controlled axes and payload between 11 and 50 lbs (5 to 23 kg).

The minimum payload required was estimated as follows:

1. The weight of the heaviest part is 1.1 lb, and a potential simultaneous loading of four parts is required.
2. The estimated weight of a gripper with several sets of jaws is 6.5 lb.

In selecting the robot, a computerized robot data bank, similar in structure to Table 9-1, was used. The six robots listed in Table 9-2 were found as appropriate candidates. All four vendors in the table have a record of hundreds of robot installations and provide service and spare parts on short notice. The breakdown of the robot system prices in

TABLE 9-2 Suitable Robots for the Case Study

Robot model	Vendor	Drive	Axes	Payload, lb	System price, $
1	A	Pneumatic	4	20	35,500
2	B	Pneumatic	5	33	47,500
3	C	Hydraulic	5	50	48,000
4	C	Electric	6	22	82,000
5	B	Electric	6	13	75,500
6	D	Electric	6	14	75,500

Table 9-2 is given in Table 9-3. The additional costs in Table 9-3 are $3500 for a riser (stand), and $9500 and $10,000 for engineering and riser.

Another point to be considered is the loading-cycle time. There are three possible methods of loading the parts:

1. Loading the four parts simultaneously by a gripper with four sets of jaws. The maximum required loading-cycle time is 24 s. Only one programmed motion is required at the loading stage.

2. Loading two parts at a time by a double-jawed gripper; a pitch motion of the wrist is a necessity. The required cycle time is 12 s.

3. Loading one part by a simple gripper. The four parts are loaded by four separate programmed motions of the robot. The maximum cycle time of the robot is 6 s.

The price of the appropriate gripper depends on the loading method. The costs of grippers in Table 9-3 are given for a two-part gripper. Vendor C provides a one-part gripper for $3500, a two-part gripper for $5500, and a four-part gripper for $11,000. These numbers are used for comparison. Since the parts are supplied in a tub, the add-

TABLE 9-3 Breakdown of Robot System Cost (in Dollars)

Model	Robot price	Sensor cost	Two-part gripper cost	Additional cost	Total robot cost
1	22,000	2000	8,000	3,500	35,500
2	32,000	2000	10,000	3,500	47,500
3	31,000	2000	5,500	9,500	48,000
4	65,000	2000	5,500	9,500	82,000
5	60,000	2000	10,000	3,500	75,500
6	56,000	2000	7,500	10,000	75,500

ing of a feeder to provide oriented parts is a necessity. The prices of the feeders also depend on the loading method as follows: one-path feeder, \$33,000; two-path feeder, \$38,000; and four-path feeder, \$51,500. Therefore the extra cost for installation of two-part gripper and feeder is \$7000, and for a four-part gripper and feeder is \$26,000.

The programming capability and the cycle time of the six robots were checked, and the following conclusions were drawn:

1. Robot 1 can load only four parts at a time because of its programming capability (limited-sequence type). This requires the most expensive gripper and feeder.

2. Robots 2 and 3 cannot operate in a cycle time of 6 s. Therefore they can load simultaneously either two or four parts.

3. The electric-driven robots can operate in any loading method. Note, however, that the price difference between robots in this group and the previous group (robots 2 and 3) is \$30,000 on the average.

Based on these considerations and the robot cost, the following conclusions were made:

1. The application does not justify the use of electric-driven robots, which are much more expensive. This leaves the slower robots (1, 2, and 3) and dictates that parts must be loaded either in pairs or four at a time. Therefore, the price of the feeder and gripper is increased by \$7000 (for pairs) or \$26,000 (for quadruplet loading). This price difference, however, is still smaller than the incremental cost (approximately \$30,000) paid for the electric robots.

2. Simultaneous loading of two rather than four parts is recommended, since it reduces the complexity and cost of both the feeder and the gripper.

3. The cheapest robot, robot 1, is not recommended for this application. Its programming method requires the use of a four-part loading. However, the basic price difference between robot 1, and robots 2 and 3 (i.e., \$12,000 or \$12,500) is smaller than the cost difference between a four-part and a two-part feeder-gripper system (\$19,000), meaning that if the cost of the whole system is considered, then the use of robot 1 rather than robots 2 or 3 is not justified.

4. The two remaining robots, 2 and 3, have the required capabilities to perform the task, and their price is comparable. Robot 3 is more sophisticated than robot 2, and three of its five axes are fully programmable. Robot 3, however, is large in size, and because of space

restriction in the machine neighborhood, there is not enough room for it. Therefore robot 2 is the best choice for the application.

The next step is to conduct a cost analysis. The total cost of the robot system used in this study is as follows:

Robot (basic price)		$32,000
Accessories:	Gripper (for two parts)	10,000
	Riser	3,500
	Sensors	2,000
		$47,500
Feeder (2-path)		38,000
Installation and shipping		2,500
Total capital investment		$88,000

Note that the price of the system is almost three times the basic price of the robot. In addition to this investment, there is an annual maintenance cost estimated at 10 percent of the cost of the robot and accessories, i.e., $4750.

The robot will save two operators paid $20.4 per hour (including fringe benefits). Based upon 16 h per day and 240 days per year, the total annual savings are $78,336. According to Eq. (9-3), the payback period is

$$\frac{88,000}{78,336 - 7750} = 1.2 \text{ years}$$

which is a typical number for a two-shift operation.

Besides studying the systematic approach to achieve the goals of this robotic project, you might notice the following:

1. The basic price of the robot is $32,000, but the total installation price (excluding the feeder) is estimated as $50,000, namely 1.6 times higher. In another case study (Stauffer, 1982b), the ratio was found to be 1.7 (the robot $28,300, the total $47,800). Therefore, as a rule of thumb, multiply the basic price of the robot by 2 to obtain the total installation price in the plant.

2. If the application requires a special gripper, the vendor should be asked to design and build the gripper. The vendor has more experience, and having the vendor design the gripper also means that responsibility would not be split.

3. Based on two-shift operation the payback period for most robot installations is between 1 and 2 years.

4. During the installation phase, safety measures must be studied and

solved. Safety devices will add extra cost to the project, but safety must be a primary goal in each robot installation.

9.5 ROBOT SAFETY

In the autumn of 1981 a sensational story was published in the newspapers—a Japanese worker was stabbed to death by a robot (see Fig. 9-2). Are robots dangerous? On the one hand we say that robots take over the hazardous jobs, but if at the same time new risks are involved, do we really remove hazardous situations?

One thing that must be kept in mind is that a robot is a type of automatic machine—a very sophisticated one, but a machine nevertheless. It has to be treated with respect like any other piece of manufacturing equipment. Accidents involving robots can happen just as with other production machinery. The robot workplace must be watched to prevent operators from hurting themselves through carelessness, exactly as it is done with other automated machines. Robots, however, may need more attention, since their workspace is much larger than the occupied floor space, and people may not be aware of the danger of the moving manipulator.

A survey of robot-related accidents over a 30-month period from January 1976 to June 1978 was made in the steel industry in Sweden (Sugimoto & Kawaguchi, 1983). The survey found that the industry, which uses about 270 robots, had reported 15 accidents in the 30-month period. While the accidents were not described in detail, the survey analysis concluded that they were typical of accidents caused by robots, involving workers who carelessly got too close to the machines while they were operating, or workers who operated robots improperly by pushing the wrong control buttons, or who were cut by tools held by robots. Upon analysis of the survey, Swedish officials concluded that, while robot disasters are few, the accident rate could be considered as one accident per 45 robots per year.

Worker stabbed to death by robot

TOKYO (AP) — A 37-year-old factory maintenance worker was stabbed to death by a robot that suddenly started up and pinned him against another machine, a government report said today. It was the first recorded fatality blamed on one of the approximately 70,000 robots in use in Japanese industrial plants. The accident occurred at the Kawasaki Heavy Industries in Tokyo last July, but it was kept secret until today, after the investigation was completed.

FIG. 9-2 A story which appeared in the news media in late 1981.

An automobile assembly plant in Japan has compiled the problems it has encountered with robots (Sugimoto & Kawaguchi, 1983). At the facility, these accidents occurred:

1. In a training course using a robot, the slewing shaft suddenly swung away from its programmed direction.
2. The arm of a robot suddenly shot up as the oil-pressure source was cut off after the robot ended work.
3. A robot made a motion that was not part of its program.
4. A robot started moving as soon as its power source was switched on, although its interlock conditions were still not ready.
5. When operating alone, a robot destroyed the work it was to weld because of a mistake in program instruction.
6. During hot summer weather, the arm of a robot sprang up, although it had otherwise been working normally.

Malfunctions such as the unexpected start reportedly occur at a rate of several dozen annually at the factory. The most common causes are reported to be electrical noise, oil-pressure valve troubles, encoder-related problems, electronic malfunctions, and mistakes by human workers.

Safety measures for using robots must be imposed for the following reasons:

1. In programming, humans must enter the workspace of robots while the robots are operating.
2. Monitoring, tool changing, inspection, and other operations involving robots or their peripheral equipment still must be done by humans.
3. To correct problems with peripheral equipment, it is necessary to enter the workspace of robots.
4. Adequate reliability of robots has not been assured.
5. Since each robot installation is different, each presents unique application problems.
6. In programmed or accidental halt the operator might enter the workspace to inspect the work or investigate the trouble. An unexpected start of the robot is possible under these circumstances.

The issue of robot safety deals with three aspects of the problem: (1) the design of a reliable control system to prevent malfunctions, (2) the

FIG. 9-3 A fence with an interlocked gate is used in surrounding the robot work envelope. The robot transfers an overliner through a two-die stamping press. (Prab.)

design of the workstation layout, and (3) training of plant personnel (programmers, operators, and maintenance staff). While the first aspect depends on the robot manufacturer, the other two must be taken care of in the plant. The following guidelines can help to remove hazardous situations to robot personnel, factory workers, visitors, and to the robot itself:

1. The robot working area should be closed by permanent barriers (e.g., fences, rolls, and chains) to prevent people from entering the area while the robot is working. The robot's reach envelope diagram (see Fig. 2-11) should be used in planning the barriers. A barrier which consists of a chain-link fence is shown in Fig. 9-3. The advantage of a fence-type barrier is that is is also capable of stopping a part which might be released by the robot's gripper while in motion.

2. Access gates to the closed working area of the robot should be interlocked with the robot control. Once such a gate is opened, it automatically shuts down the robot system.

3. An illuminated working sign, stating "robot at work," should be automatically turned on when the robot is switched on. This lighted sign warns visitors not to enter into the closed area when the robot is switched on, even if it does not move.

4. Emergency stop buttons must be provided in easily accessible loca-

tions as well as on the robot's teach box and control console. Hitting the emergency button stops power to the motors and causes the brakes on each joint to be applied.

5. Pressure-sensitive pads can be put on the floor around the robot that, when stepped on, turn the robot controller off.

6. Emphasize safety practices during robot maintenance. In addition, the arm can be blocked up on a specially built holding device before any service work is started.

7. Great care must be taken during programming with the manual reaching mode (see Sec. 7.1) The teach box must be designed so that the robot can move as long as a switch is pressed by the operator's finger. Removing the finger must cease all robot motions. The motion during the teaching mode is restricted to slow speeds in some robots. Programmers should stay out of areas where they might be hurt in the event of a robot malfunction.

8. The robot's electrical and hydraulic installation should meet proper standards. This includes efficient grounding of the robot body. Electric cables must be located where they cannot be damaged by the movements of the robot. This is especially important when the robot carries electrical tools such as a spot-welding gun.

9. Power cables and signal wires must not create hazards if they are accidentally cut during the operation of the robot.

10. If a robot works in cooperation with an operator, for example when a robot forwards parts to a human assembler, the robot must be programmed to extend its arm to the maximum when forwarding the parts so that the worker can stand beyond the reach of the arm.

11. Mechanical stoppers, interlocks, and sensors can be added to limit the robot's reach envelope when the maximum range is not required. If a wall or a piece of machinery not served by the robot is located inside the reach envelope, the robot can be prevented from entering into this area by adding photoelectric devices, stoppers, or interlock switches in the appropriate spots. There are robots supplied with adjustable mechanical stoppers for this purpose (e.g., the one presented in Fig. 9-3).

In the case of the Japanese factory maintenance worker who was stabbed to death by a robot, two of these guidelines were violated. First the robot controller was not designed so that it could not unintentionally be operated. The worker put the robot on manual control for some maintenance function, but as the report indicates, he accidentally brushed against the "on" switch, causing the robot's clawlike appendage to pin him against a nearby automobile gear tooling machine. The control should have been designed differently to prevent the accidental start-up of the robot. Second, the emergency stop switch was not properly designed in that it was apparently not conspicuous, easily identifiable, or conveniently located. The accident report indicated that other workers rushed to the scene but were unable to stop the robot's action (Howard, 1982).

Another approach states that only robots themselves are able to detect the approach of humans. Therefore, the solution to the safety problem is to provide sensor systems which can detect intruders that enter the robot area while it is operating. The American National Bureau of Standards (NBS) divides the sensor systems into three levels (Kilmer, 1982):

Level I: Perimeter penetration detection around the workstation

Level II: Intruder detection within the workstation

Level III: Intruder detection very near the robot (a "safety skin").

Level I systems provide perimeter penetration detection around the robot workstation. These systems provide an indication of an intruder crossing the workstation boundary, but they do not necessarily provide any information regarding the location of the intruder within the workstation. The simplest safety strategy approach might be to use the level I system to alert personnel that they are entering a robot workstation and that they should exercise extreme caution or to provide a preliminary signal to the robot control system to check the status of other safety sensors.

Level II systems provide detection in the region between the workstation perimeter and some point on, or just inside, the working volume of the robot. The actual boundaries of this region are dependent upon the workstation layout and the safety strategy being employed for a particular robot design. In some cases, it may be permissible for personnel to be inside the workstation and perhaps even inside a portion of the accessible working volume of the robot while the robot is operating. In others, it may be necessary to slow down or halt all robot movements as soon as an intruder gets within a specified distance of the robot.

Level III systems provide detection within the robot working volume. This type of system, sometimes referred to as a safety skin, is required for cases where personnel must work close to the robot, such as during teach-mode operations. In such cases, the robot must be operational even though someone is within the working volume. The level

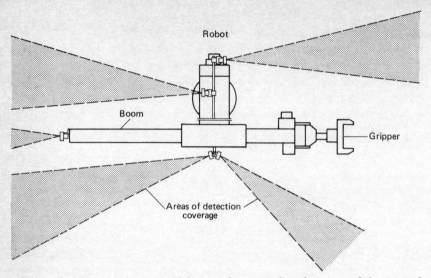

FIG. 9-4 Approximate locations of the transducers on the robot arm and the areas of detection coverage.

III system must be capable of sensing and avoiding an imminent collision between the robot and the operator in the event of some unexpected movement. Because the distance between the robot and the operator is much less in this case, the response time of the level III safety system must be much shorter than for the level I or II systems. These smaller separation distances also impose a requirement for finer distance-resolving capabilities in the level III system.

The three-level safety system was applied in the NBS robot laboratory to protect the work area of a "Stanford Arm" robot (Kilmer, 1982). The transducers are ultrasonic echo-ranging sensors with a $\pm 10°$ cone in which the intruder can be detected. The transducers were positioned to provide coverage in the areas illustrated in Fig. 9-4. The robot gripper and the end of the boom are the two areas of primary coverage, because these locations have the highest potential for a collision. There is no coverage to the sides. The only way the robot can strike an intruder is by rotating to the right or left. However, before a collision could occur, the intruder would be detected by one of the transducers, since the cone sweeps across the intruder's location before the robot's arm does. The other area not covered is directly in front of the robot gripper. Although this is not desirable and will be eliminated in more advanced safety system design, this is not a problem in the NBS's experiments because the gripper is always operating over the table top, which extends beyond the reach of the robot arm.

Since robot installations are not standard and depend on the exact location in the plant, it is difficult to provide guidelines or a design of a sensor safety system which will fit all cases. Therefore, it is wise to have a safety engineer check out the installation before putting the robot to production work.

REFERENCES

Behuniak, J. A.: "Planning the Successful Robot Installation," *Robot. Today*, vol. 3, no. 2, Summer 1981, pp. 36–37.

Bergant, J. M.: "Achieving a Successful First Robot Installation," *Robot.Today*, vol. 4, no. 1, February 1982, pp. 63–66.

Engelberger, J.: *Robotics in Practice*, AMACOM, New York, 1980.

Howard, J. M.: "Human Factors Issues in the Factory Integration of Robotics," *Robots 6*, MS82-127, March 1982. Also published in *Robot. Today*, vol. 4, no. 6, December 1982, pp. 32–34.

Janjua, M. S.: "Selection of a Manufacturing Process for Robots," *Ind. Robot*, vol. 9, no. 2, June 1982, pp. 97–101.

Kilmer, R. D.: "Safety Sensor Systems for Industrial Robots," *Proc. Robots 6*, March 1982, pp. 479–491.

Marci, G. C.: "Analysis of First UTD Installation Failures," SME Technical Paper, MS77-735, 1977.

Naidish, N. L.: "Return on Robots," *Robots 6*, Paper MS-82-136, March 1982.

Ottinger, L. V.: "A Plant Search for Possible Robot Applications," *Ind. Eng.*, vol. 13, no. 12, December 1981, pp. 26–32.

Ottinger, L. V.: "Evaluating Potential Robot Applications in a System Context," *Ind. Eng.*, vol. 14, no. 1, January 1982, pp. 80–86.

Stauffer, R. N.: "Anatomy of a Successful Robot Arc Welding Installation," *Robot. Today*, vol. 4, no. 2, April 1982a, pp. 41–42.

Stauffer, R. N.: "Automating an Injection Molding Operation," *Robotics Today*, vol. 4, no. 4, August 1982b, pp. 23–27.

Sugimoto, N., and K. Kawaguchi: "Fault Tree Analysis of Hazards Created by Robots," *13th Int. Symp. Ind. Robots and Robots 7*, vol. 1, April 1983, pp. 9–13.

CHAPTER 10

Computer-Integrated Manufacturing Systems

A quiet revolution is going on in the manufacturing world that is changing the look of the factory. Computers are controlling machines and robots are performing processes, and both are doing it far more efficiently than human operators. The high degree of automation, that until recently was reserved for mass production only, is applied now, with the aid of robots and computers, also to small batches. This requires a change from hard automation in the production line to a flexible manufacturing system which can be more readily rearranged to handle new market requirements.

Flexible manufacturing systems combined with automatic assembly and product inspection on one hand, and CAD/CAM systems on the other, are the basic components of the factory of the future. The supervision of this factory, which is predicted to be in place by the end of the twentieth century, will be performed by computer-integrated manufacturing system, in which the production flow, from the conceptual design through the finished product, will be entirely under computer control and management.

10.1 HIERARCHICAL COMPUTER CONTROL

The availability of computers in the manufacturing plant has brought a hierarchical structure of computer control to the factory, as illustrated in Fig. 10-1. The lowest level of this structure contains stand-alone

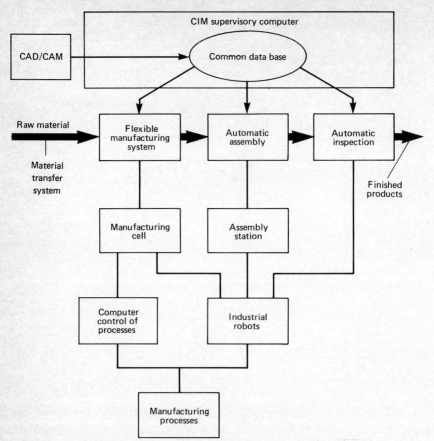

FIG. 10-1   Hierarchical structure of computers in the factory of the future.

computer control systems of manufacturing processes and industrial robots. The computer control of processes includes all types of CNC machine tools, welders, EDM, ECM, and high-power lasers as well as the adaptive control of these processes.

The operation of several CNC machines and a single robot can be incorporated into a system which produces a specific part, or several parts with similar geometry. This structure is denoted as the *manufacturing cell* (see Sec. 6.2). The computer of the cell is interfaced with the computers of the robot and the CNC machines. It receives "completion of job" signals from the machines and issues instructions to the robot to load and unload the machines and change their tools. The software includes strategies permitting the handling of machine breakdown, tool breakage, and other special situations.

The operation of many manufacturing cells can be coordinated by a central computer via the aid of a material-handling system. This is the highest hierachical level in computer control of a manufacturing plant and is denoted as a *flexible manufacturing system (FMS)*. The FMS accepts incoming workpieces and processes them, under computer control, into finished parts.

The parts produced by the FMS must be assembled into a final product. In the factory of the future the parts will be routed on a transfer system to assembly stations. In each station, a robot will assemble parts either into a subassembly or (for simple units) into the final product (see Sec. 6.5). The subassemblies will be further assembled by robots located in other stations. The final product will be tested by an automatic inspection system.

Another type of computer utilization is the CAD/CAM system. The FMS uses CAD/CAM systems to integrate the design and manufacturing of parts in order to minimize the production cycle in the factory. A criterion for design can be the product's suitability for robot assembly. In this case the CAD/CAM computer must have data associated with the assembly plant.

At the highest hierachical level there will be *computer-integrated manufacturing* (CIM) systems. Such systems call for the coordinated participation of computers in all phases of a manufacturing enterprise: the design of the product, the planning of its manufacture, the automatic production of parts, automatic assembly, automatic testing, and, of course, the computer-controlled flow of materials and parts through the plant. All these phases must be integrated into one computer network, supervised by the CIM central computer that monitors the interrelated tasks and controls each of them based on an overall management strategy.

The development of CIM systems is possible because of the improving capability of computer technology, coupled with the pressure for higher productivity, a combination that is resulting in a new era in manufacturing. While robots and CNC manufacturing systems are replacing the human's *power and skill,* CIM systems will be replacing the human's *intelligence,* and using it with incomparably higher efficiency.

The increase in productivity associated with CIM systems will not come about from the speeding up of the machining operations. The breakthrough in productivity is due to the concept of computer managing an integrated manufacturing system. The CIM computer will be able to make decisions and adapt the production flow to variations in the environment. For example, if a specific product fails in the final inspection station, another, similar product must be automatically manufactured in order to meet the requested quantity of output products. The CIM computer will issue an instruction to the manufacturing system to produce the additional corresponding parts required for the assembly of the product. The major increase in productivity associated with CIM systems will be achieved by minimizing the direct labor employed in the plant. In addition to the direct savings in labor costs, one can expect substantial savings from reduced inventories. Far fewer parts, whether finished or in process, will be "waiting," either for the next operation or for assembly. The anticipated result is an overall cost reduction by a factor of from 5 to 10, in other words, a cost reduction of from 80 to 90 percent (Cook, 1975).

The remainder of this chapter is mainly devoted to the description of some of the components of a CIM system: The FMS and CAD/CAM which together draw the outline of the factory of the future.

10.2 FLEXIBLE MANUFACTURING SYSTEMS

Various individual manufacturing systems can be incorporated into a single large-scale system in which the production of parts is controlled with the aid of a central computer. The advantage of such a production system is its high flexibility in terms of the small effort and short time required to manufacture a new product, and therefore it is denoted as a *flexible manufacturing system* (FMS). It is the flexibility of the robot which enables its use as a major component in FMSs.

10.2.1 The FMS Concept

The FMS provides the efficiency of mass production for batch production. The term "batch production" is applied to parts manufactured in lots ranging from several units to more than 50, for which the total annual demand is fewer than, say, 100,000 units. The term "mass production" applies when higher annual production rates are required and the use of special-purpose machines can be justified. To machine a single unit with general-purpose machine tools may cost 100 times as much as to manufacture the same part by the most efficient mass-production methods (Cook, 1975). As an example, consider a complex mass-produced part with which almost everyone is familiar: the cylinder block for a typical V-8 automobile engine. Under mass-production conditions, where the engine block is conveyed automatically along a transfer line, with the various operations (drilling, tapping, boring, milling) being executed in sequence at the different stations along the line, the complete machining cost (excluding the raw material) would be on the order of $25. If, however, only a few special cylinder blocks were to be made with machine tools and skilled labor, the machining cost per block could easily rise from $25 to $2,500 or more (Cook,

1975).† By making use of FMS technology it should be possible to reduce the cost of producing parts in small and medium quantities.

Existing flexible machining systems in the United States are typically made up of machining centers working in concert with other types of machines, all under the control of a central computer. The workpieces are on pallets which move throughout the system, transferred by wire-guided carts or towlines located beneath the floor or by some other mechanism. These flexible machining systems limit handling by the operators and can be more readily reprogrammed to handle new requirements.

Future FMSs will contain many manufacturing cells, each cell consists of a robot serving several CNC machine tools (see Sec. 6.2) or other stand-alone systems such as welders, EDM machines, etc. The manufacturing cells will be located along a central material transfer system, such as a conveyor, on which a variety of different workpieces and parts are moving. The production of each part will require processing through a different combination of manufacturing cells. In many cases more than one cell can perform a given processing step. When a specific workpiece approaches the required cell on the conveyor, the corresponding robot will pick it and load it into a CNC machine in the cell. After processing in that cell, the robot will return the semifinished or finished part to the conveyor. A semifinished part will move on the conveyor until it approaches a subsequent cell in which its processing can be continued. The corresponding robot will pick it up and load it into the machine. This sequence will be repeated along the conveyor, until, at the end of the route, there will be only finished parts moving. Then they can be routed to an automatic inspection station and subsequently unloaded from the FMS. The coordination among the manufacturing cells and the control of the parts flow on the conveyor will be accomplished under the supervision of the central computer.

Advanced FMSs will contain a high-power laser station incorporated in the production line. The laser will be used mainly for heat treatment, sheet-metal cutting, drilling, and welding. At present, laser treatment of materials with CO_2 lasers in the 5- to 15-kW range are becoming more popular in industry. The central computer of advanced FMSs will contain a machining data base to provide the recommended cutting parameters to the machine tools in the plant, based upon a selected tool, workpiece material, and maximization of the production rate in the entire plant.

In the early eighties, there has been a lack of clarity in definitions related to FMS technology (Drozda, 1983). In machine tool exhibitions many pieces of manufacturing equipment were promoted as FMSs. In

---

†Although the absolute costs have been changed since 1975, the cost ratio is still valid.

some cases machine tool builders merely added a robot loading arm to an existing lathe and then entered the market ready to sell FMSs. While such and other stand-alone machines provide more flexibility, they cannot be defined as FMSs.

A true FMS installation must include three elements:

1. Workstations, such as manufacturing (or machining) cells, head changers, machining centers, assembly stations, and welding stations.
2. Material transfer system, such as unmanned carriers (towline or wire-guided carts) or conveyor systems (roller or belt type).
3. Central computer that controls the flow of materials, tools, and information (e.g., machining data and machine malfunctions) throughout the system. The software associated with the control computer is an important ingredient of the FMS.

The advantages of FMS include the following:

1. Increased productivity.
2. Shorter preparation time for new products
3. Reduction of inventory parts in the plant
4. Saving labor cost
5. Improved product quality

Additional economic savings may be from such things as the operators' personal tools and gloves. Other savings are in locker rooms, showers, and cafeteria facilities. All represent valuable plant space, which will not require enlarging if company growth is achieved with flexible automation systems.

10.2.2 Transfer Systems

FMS and assembly systems utilize material transfer systems on which the workpieces, finished parts, or tools are moving between the machining centers, the manufacturing cells, or assembly stations. Various types of transfer systems have been utilized in existing FMSs and a few of them are discussed below.

Towline System. In this system workpieces are attached to pallet fixtures or platforms which are carried on carts that are towed by a chain located beneath the floor. The pallet fixture is designed so that it

FIG. 10-2   Material transfer by wire-guided carts. (Cincinnati Milacron.)

can be conveniently moved and clamped at machines in successive manufacturing cells. The advantage of this method is that the part is accurately located in the pallet, and therefore it is correctly positioned for each machining operation.

Wire-Guided Carts. Two carts, which are moving along paths determined by wire embedded in the floor, are shown in Fig. 10-2. In the foreground, the cart has picked up a finished palletized workpiece from the machining center in the background and delivers it to an unloading station elsewhere in the system. The cart in the background is about to deliver a rack of replacement tools to the receiver station on the machining center in the foreground. Tool interchange from this rack to the machining center's storage chain is automatic and computer controlled.

Roller Conveyor System. In this system a conveyor consisting of rotating rollers runs through the factory. The conveyor can transport palletized workpieces or parts which are moving at constant speed between the manufacturing cells. When a workpiece approaches the required cell it can be picked up by the robot or routed to the cell via a cross-roller conveyor. The rollers can be powered either by a chain drive or by a moving belt which provides the rotation of the rollers by friction.

Belt Conveyor System.   In this type of transfer system either a steel belt or a chain driven by pulleys is used to transfer the parts. This general type of transfer system can operate in three different methods (Grover, 1980):

1. *Continuous Transfer.*   In this type the workpieces are moving continuously and the processing is either performed during the motion or the cell's robot picks up the workpiece when it approaches the cell. The in-motion processing transfer system is used in automobile assembly lines in which human work and robot operation are combined. Figure 10-3 illustrates the operation of spot-welding robots in car assembly, where the welding is performed while the cars are moving on the transfer line. The speed of the transfer line is taken into account by the robot controller when the welding head is moved from one point to the next.

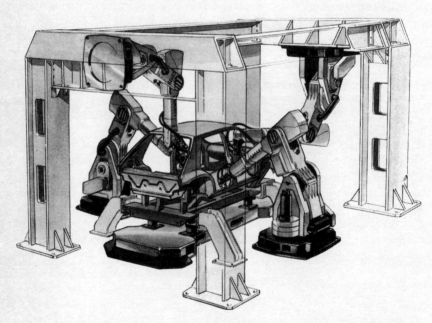

FIG. 10-3   Robot spot welding is performed during the motion of an automobile assembly line. (Coman.)

2. *Synchronous Transfer.*   This method is mainly used in automatic assembly lines. The assembly stations are located with the same distance between them and the parts to be assembled are positioned at equal distances along the conveyor. In each station a few parts are assembled by a robot or automatic device with fixed motions. The conveyor is of an indexing type, namely, it moves a short distance and stops when the product is in the stations, and subsequently the assembly takes place simultaneously in all stations. This method can be applied where station cycle times are almost common.

3. *Power-and-Free.*   This method allows each workpiece to move idependently to the next manufacturing cell for processing. Usually the method is used for large workpieces and when the manufacturing stations have varying cycle times. The main overhead conveyor loop in Fig. 10-4 is of the power-and-free type. In this case it is used to transfer heavy machining heads between machining stations.

10.2.3 Head-changing FMS

One of the operating FMSs is the Bendix Flexchanger (Stauffer, 1981*a*), a head-changing system, which is illustrated in Fig. 10-4. It consists of 3 machining stations, 77 machining heads, and 11 probe heads. Parts produced are cast-iron transmission cases and reverser cases, each weighing approximately 200 lb (90 kg).

In the Flexchanger system, the parts and fixtures remain at the machining stations, and the machining heads are transferred to and through the stations as required so that work can be performed in any combination. Fifty heads are used when running transmission cases and 38 heads are required for the reverse cases.

Operations at each machining station are controlled by a CNC unit. A programmable controller incorporated in the CNC is used to control and interface conveyor loop operation with the machining cycle. Operations performed include drilling, tapping, spot facing, reaming, rough boring, finish milling, grooving, and probing. Tools in machining heads range from a single-spindle mill to a multiple-spindle head incorporating 23 different tools. The probe heads check for broken tools and include an air blow-off to clean out loose chips and other foreign material.

Heads that are not in use at any of the three machining stations are stored in a main overhead conveyor loop, of the power-and-free type. Heads not immediately required for machining are carried in a second overhead storage loop, 18 ft (6 m) off the floor. Each loop can hold up

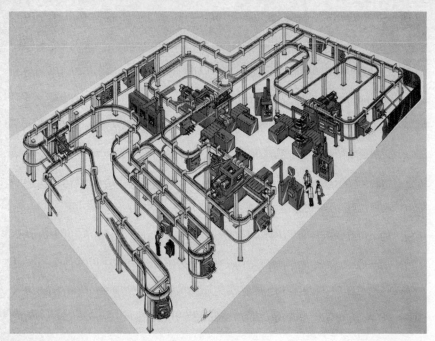

FIG. 10-4   Head-changing FMS. (Bendix.)

to 55 machining heads for a total of 110. Total length of both loops is about 1400 ft (430 m).

In addition to the two conveyor loops, the head transport system includes a standard roll-type transfer bar at each machining station. As a machining head approaches a station where it is to machine a part, the head is automatically disengaged from the conveyor chain drive mechanism. The transfer bar then simultaneously moves this head into position and the completed head moves out and back into the conveyor loop.

The incoming head is accurately located and automatically clamped to the feed unit at the machining station. Four locating pads used for positioning incorporate air blow-off to clean chips and other debris from the mounting surface. The air blow-off also serves as a signal to confirm that the head is properly mounted against the locating pads. Then the head is advanced toward the part and the machining begins. After completing its machining operations, the head is backed away from the part and transferred to the conveyor loop.

Changeover from one part to another is done sequentially. As the last part in a run is finished at each machining station, the fixture and the

CNC program are changed. Machining heads for that part are routed into storage on the conveyor, and heads for the other part are brought on stream out of storage.

The complete system includes a qualifying machine and two finishing machines—one for each type of part. The qualifier mills several locating surfaces and bores two locating holes. Finish bores and finish mill surfaces are completed in the finishing machines.

10.2.4 Variable-Mission Manufacturing System

An FMS denoted as a variable-mission manufacturing (VMM) system was developed by Cincinnati Milacron and is in operation in one of its plants.

The VMM system's basic components, as seen in Fig. 10-5, include two CIM-Xchanger 20 HC CNC machining centers that are supplied continuously with palletized parts and replacement tools by remotely controlled wire-guided carts. The carts, following electrical signals

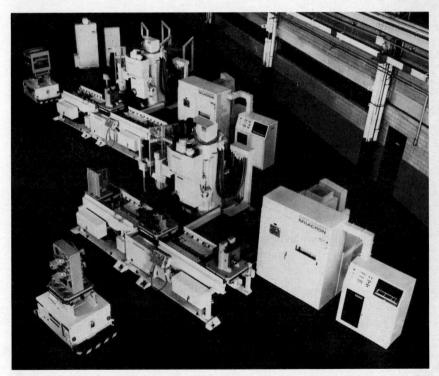

FIG. 10-5   Variable-mission manufacturing system. (Cincinnati Milacron.)

from wire embedded in the floor, move along a closed-loop layout between and around the two machining centers, as well as into and out of pallet loading and unloading stations. The cart movement and the entire system management are under control of two computers.

Seven different pallets move through the system at all times. Five of these pallets are set up to carry different production parts. They move from load stations through the layout to a pallet delivery station at one or the other of the machining centers. The parts then move to the machine's rotary index work module and are clamped automatically in position for machining. Following machining, each part is moved from the work module to a pallet discharge station where it is picked up by one of the carts and moved through the system to one of two load-unload stations.

Two of the seven pallets carry a tool rack holding twelve tools. These are delivered intermittently to each of the machining centers, where automatic tool interchange is programmed and tools are changed between the rack and the machine's storage chain. This tool change could be made during a system's unmanned second- or third-shift operation. Special features of the system include utilization of a surface-sensing probe which is stored in the tool chain like any other tool and brought to the spindle nose on command from the control. The probe can identify true work position and/or other part-shape anomalies (such as casting variations). It can feed information back to the computer control of the system to correct for differences or errors. It can detect incomplete machining operations caused by broken tools or worn tools and can detect missing surfaces.

The VMM system includes an optional adaptive control feature called *torque-controlled machining.* Microprocessor based, this feature increases productivity by acting as an adaptive control of feedrate, which depends on torque at the tool and horsepower at the spindle motor. It can sense such conditions as air gaps, workpiece hardness, hole breakthrough, cross and blind holes, and chip clogging.

The wire-guided carts offer significant advantages in flexibility in that it is relatively simple to add, remove, and/or reroute carts at any time. The carts can carry up to 4000 lb. They are battery-powered, with sufficient energy-storage capability to easily serve an 8-hr shift.

The overall VMM control system has been designed in a modular fashion, with individual modules responsible for separate functions. The modules provided are these:

Data Distributor: Sends data to and receives data from the machine tool control. The data are normally in the form of NC part programs but can also be tooling data or operator messages.

Data Manager: Stores and provides the capability for the user to manage the data in the system. It is also responsible for data transmission to and from the user's remote computer.

Traffic Coordinator: Controls workpiece movement between workstations within the VMM system.

Work Preparation: Guides the VMM operator in the activities required prior to entering parts into the system. These activities include work-order definition, determination of lot sizes, and the entry of start-stop dates.

Tool Manager: Stores, monitors, and updates tool data files which contain tool length and diameter compensation, tool life expectancy, etc.

The control of the VMM system is essentially two-level. The two machining centers are interfaced with their usual CNC consoles which control the typical machine functions, such as positioning and machining cycles. The CNCs receive their data via computer, however, and not from punched tape. On another level is the overall management of the VMM system. The hardware for this level includes two computers, video display terminal, two disk drives, and hard copy console. The computers are a PDP-11/34 which provides overall control of the system, and a PDP-11/35 which provides control over the cart movement. In addition, there is a shop terminal interfaced with the PDP-11/35. This terminal is used primarily to enter commands in a semiautomatic mode. When memory is added to this unit, it can be used as a "behind-the-tape reader" interface, which serves like hard-wired machine controls to the VMM system (Koren, 1983).

10.3 FMSs IN JAPAN

The sales of robots in Japan amounted to $400 million in 1981 (Weill, 1983), compared with less than $200 million in the United States in 1982. Forecasts predict an output of $2.5 billion in 1990 (Weill, 1983), of which about $2 billion are sales of industrial robots[†] as stand-alone units or incorporated in FMSs, and the other $500 million are for robots in nonmanufacturing sectors, such as agriculture, ocean resources, nuclear plants, medicine, and construction.

The major effort in advanced manufacturing in Japan is toward the system approach, namely the integration of several technologies

†*Business Week*, February 9, 1981, p.64, predicted robot sales of $2 billion in the United States by 1990.

including robotics into a comprehensive, flexible system. In the 11th Japan International Machine Tool Fair held in October 1982, 25 FMSs were on display, and it was estimated that at that time there were about 100 FMSs in operation in Japan (Weill, 1983).

Japanese FMSs are characterized by three elements: workstations (manufacturing cells, machining centers, or assembly stations), unmanned carriers for material handling, and a central supervisory computer (Iwata, 1982).

The systems are using a variety of sensors to monitor the machine production status. Common sensors are tool breakage detectors (most of them are based on spindle motor current), and tool wear monitors (based on axial motor current, acoustic emission, etc.). The sensor outputs are automatically fed to the central computer. Many systems use automatic tool-changing schedules based on programmed tool life estimations.

A description of several FMS installations is given below.

10.3.1 FANUC's Fuji Complex

The FANUC Company has established a complex consisting of two plants in Oshino (located in the foothills of Mount Fuji) which represents one of the most advanced examples of implementation of the FMS concept. The FANUC's Fuji complex, which specializes in the production of robots and computer-controlled machine tools, was established in 1980. Operating around the clock, it accounts for a major portion of the world's robot production.

The two plants accommodate the Motor Manufacturing Division, which produces dc motors for robots and CNCs, and the Mechatronics Manufacturing Division that produces robots and computerized machine tools. ("Mechatronics" refers to a combination of mechanical and electronic technologies.)

In the Motor Manufacturing Division there are approximately 60 employees using 100 robots to produce 40 different kinds of dc servomotors at a rate of 10,000 units per month. This division is considered the world's most advanced computer-controlled "unmanned" factory. This first floor of the plant is dedicated to parts machining, whereas the second floor is occupied by assembly operations.

The first floor contains 60 manufacturing (or machining) cells with robots and delivery tables, as shown schematically in Fig. 10-6. Machining is carried out for some 900 varieties of motor parts, in lots ranging from 20 to 1000 units. Two unmanned carriers transport the parts, with all transfer of workpieces being handled by robots. In the

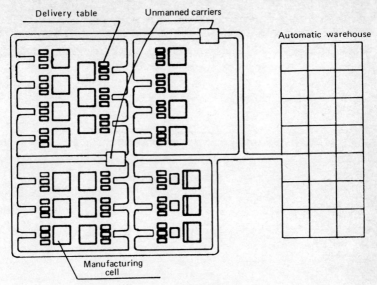

FIG. 10-6   Schematic structure of FANUC's FMS.

central control room, a special scheduling method adjusts the process planning according to production plans coming from the head office.

Parts machined on the first floor are temporarily stored in the automatic warehouse and retrieved when needed for automatic assembly completion. The assembly stations, or cells, each incorporating one or more robots, are arranged in linear configurations and served by unmanned carriers which transport pallets. The cell robot picks up the parts from the pallet (which was left by the carrier), and assembly is performed. When the assembly work allocated to a given cell is completed, the subassembled product is placed on a workpiece feeder and sent to the next assembly station.

In the Mechatronics Manufacturing Division a permanent staff of 100 employees (30 in machining and 70 in assembly, an operation which remains mainly manual) produces 300 robots, 100 wire-cutting electrodischarge machines, and 100 CNC machine tools every month. Materials delivered to the plant are first stored in a special automatic warehouse. When needed, they are transported to the manufacturing cells using three unmanned carriers. Machining is performed and the completed workpieces are sent to the parts warehouse for temporary storage. As many as 450 different types of parts can be processed in lots of from 5 to 20 units.

The production is essentially based on FANUC's factory automation

FIG. 10-7   FMS consists of manufacturing cells. (FANUC.)

(FA) systems which are composed primarily of manufacturing cells (which consist of CNC machine tools with robots or pallet changers) and assembly cells (which consist of an assembly robot and peripheral equipment). The FA units can be combined to form an FMS, as shown in Fig 10-7. In the FMS, unmanned carriers are used to transport workpieces between cells, and an automated warehouse is incorporated for storing workpieces, all under the control of a host computer.

The Mechatronics Manufacturing Division works in three shifts, of which two are unmanned. The Motor Manufacturing Division works in three unmanned shifts in assembly, which is slower than machining. The material for the second and third shifts is prepared during the first shift, on pallet pools. During the first shift, loading and setting of parts are carried out locally by operators according to a daily plan. The tool stockers are also filled up in the first shift. Before tool wear or breakage can occur, a monitoring system based on spindle current automatically performs tool changes. Individual manufacturing cells can be disconnected in case of malfunction without stopping the whole manufacturing process.

What makes the FANUC plant so special is that the whole process comes within a hair of the science-fiction fantasy of "robots making robots." At the Fuji complex FANUC boasts a unique FMS, integrating computerized machine tools, assembly robots, automatic delivery carriers, and an automated warehouse with central computer control of the whole factory.

10.3.2 The Yamazaki FMS

In its factory in Oguchi (Nagoya), the Yamazaki Machinery Works Ltd. has installed a powerful FMS comprising 10 machining centers and a cart transportation system under the control of a central computer. The investment for the equipment amounted to $18 million. The system is served by two shifts of six operators each, and a night shift of a single operator. This reduced labor power replaces a conventional team of 215 persons. Work material is prepared by the operators on the carts which transport the workpieces to the machining centers.

The setting of the tools is carried out in a tool room. The tools are stored in special huge drums which are moved to the machines for tool stocker change by a crane. Tool compensations are given to the machine control units. An effort at standardization of tools was made in order to reduce the number of tools (about 60) and to optimize the tool change time, which should be the same for all tools on a tool drum.

The plant comprises two lines. On one line, the workpieces are headstocks on CNC lathes or machining centers; they can be 3.3 ft (1 m) in length and weigh up to 3000 kg. On the other line, heavier workpieces, like machine tool beds or columns weighing up to 8000 kg, can be handled. The overall machining capacity is 600 workpieces per month. In the computer room, 74 types of schedule programs have been stored for 1400 workpieces.

Compared with a conventional system, Yamazaki's FMS needs only 13 persons instead of the 215 in a conventional installation (a reduction of 18:1). It shortens the flow time from 3 months to 3 days, and the number of workpieces in production is reduced from 2760 parts to 120 parts (ratio 23:1). Therefore the personnel problems are less critical in case of a recession, and the system can quickly react to an increase of orders.

The comparison between a conventional system and the FMS system (for 1 year) is summarized in Table 10-1 (Weill, 1983), which shows

TABLE 10-1 A Cost Comparison between Conventional and FMS Systems (in $1000)

Expenses	Conventional	FMS
Investment in plant	14,088	18,177
In-processing inventory	5,018	218
Production	20,700	27,310
Manufacturing costs	21,591	22,963
Profit before tax	−(891)	4,347
Labor expense	3,960 (215 workers)	227 (13 workers)

that the FMS profit comes from lower labor expenses, lower in-process inventory, and higher production volume.

10.3.3 Okuma's FMS

Okuma Machinery Works Co., situated in Oguchi, has installed a comprehensive FMS in its own plant. It consists of seven identical machining centers served by unmanned carriers, called Robotrailers. The carriers can travel along a fixed track [525 ft (160 m) long] and have 47 stop points. Materials fixed on pallets are transported by the carriers either directly to the machining centers or to pallet pool stations when running machines at night without operaters. Machined workpieces are automatically returned by the carriers. The carriers have access to the tool control room, where they are loaded with tools set on pallets to be transported to magazine stations beside the machines.

The whole FMS operation is controlled by a supervisory computer built in-house (Okuma Campus 5000). Machining data, according to a schedule, are transmitted from the computer to the machining centers through optical-fiber cables. Since the system uses identical machining centers, parts can be produced on any machine depending upon the schedule set by the computer, thus enhancing the system flexibility.

Compared with conventional operation, which requires five operators per shift, the FMS requires only four operators for 24-hour-a-day operation. To guarantee the production rate and part quality, the system is equipped with a variety of sensors, such as tool detectors and dimension-gauging sensors, as well as automatic tool-length compensation and chip-removal conveyor systems.

10.3.4 Conclusions

The Japanese record shows that flexible manufacturing technology, when properly applied, can indeed fulfill the promises of boosting productivity and lowering cost in batch production. From the user's viewpoint, however, the justification of higher cost involved with FMS is no easy matter. When considering the introduction of FMS technology, vendor selection becomes very critical, and it is most important to pick a vendor that has a good installation record and is committed to FMS technology.

10.4 CAD/CAM SYSTEMS

Computer-aided design (CAD) means the use of a computer to assist in the design of an individual part or a system, such as an aircraft. The

CAD process involves two basic steps: the design of a model with computer graphics and computer analysis of the model. Many CAD systems also include kinematics programs for animating motion of robot manipulators and other mechanisms.

Computer-aided manufacturing (CAM) means the use of a computer to assist in the manufacture of a part. CAM can be divided into two main classes: (1) On-line applications, namely the use of the computer to control robots and manufacturing systems in real-time, such as CNC and adaptive control systems of machine tools; (2) off-line applications, namely the use of the computer in process planning and non-real-time assistance to the manufacturing of parts. Examples of off-line CAM are the preparation of part programs in CNC or task programs in robotics.

CAD/CAM is a unified software system, in which the CAD portion is interfaced inside the computer with the CAM one. A complete layout of a CAD/CAM system and its environment are presented in Fig 10-8. Double-line blocks in Fig 10-8 represent hardware which is not part of the CAD/CAM system but is related to it.

The main concept of CAD/CAM systems is the generation of a common geometric data base which is used for all the design and manufacturing activities. These include specifications of the product, conceptual design, final design, drafting, manufacturing, assembly, and inspection (computer-aided testing). At each stage of this process, data can be added, modified, used, and distributed over networks of terminals and computers. The single data base provides a substantial reduction in human errors and a significant shortening of the time required from the introduction of a concept of a product to the manufacturing of the final physical product.

The size and capability of the required computer system depends on the complexity of the product. In the aerospace and aeronautics industry, where a complete aircraft can be designed with a CAD/CAM processor, the system must accommodate new data and changes in data arriving from a variety of users. In this industry, hundreds of engineers have access through interactive graphic terminals to the same huge data base, which includes the geometry of thousands of aircraft components. Therefore, these systems must have a strong data base management capability. By contrast, if simple products are designed by a company, the required CAD/CAM system would need only one computer terminal. Today the major users of CAD/CAM systems are major aerospace and automotive industries, but the declining prices of these systems rapidly enlarges the number of other users.

The end result of current CAD/CAM systems is usually engineering drawings or a part program in the form of a list or a punched tape. In advanced CAD/CAM systems part and task programs can be directly fed into the control computers of CNC machines and robot systems.

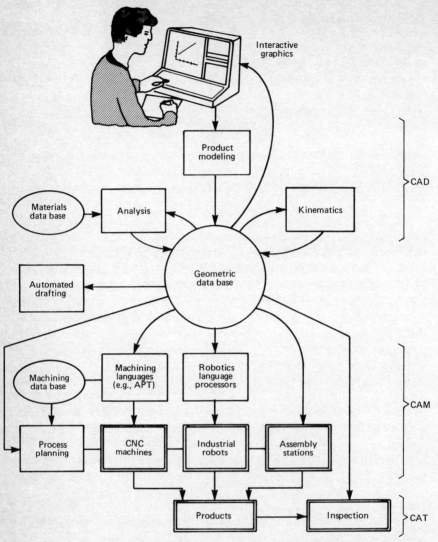

FIG. 10-8 Layout of a complete CAD/CAM system. Double-line blocks represent hardware interfaced with CAD/CAM.

10.4.1 Computer-Aided Design

A CAD system is basically a design tool in which the computer is used to analyze various aspects of a designed product. The CAD system supports the design process at all levels—conceptual, preliminary, and final design.

There are many areas in which CAD has made significant contributions, such as architecture, civil engineering, transportation, cartography, and energy-resource analysis. However, the major impact of CAD is in the following applications:

1. Modeling of complex three-dimensional parts and the assembly of parts into a product, with emphasis on reducing design iterations required to achieve desired product characteristics.
2. Design of aircrafts and automobiles. For example, Boeing's 757, 767, and 777 models were designed with a CAD/CAM system.
3. Microelectronics mask layout, where the job cannot be done manually.
4. Packaging of dense printed-circuit boards for electronic assemblies.
5. Automated drafting for large industries with hundreds of drafters.

In addition, the CAD's software interface with the product's manufacturing, assembly, and inspection has a significant impact since it reduces the time from a conceptual design to the introduction of the product and elevates the quality of the product.

The CAD process may be divided into three major stages as shown in Fig. 10-8: Modeling, engineering analysis, and kinematics, which is used in the design of moving mechanisms.

Modeling. The designer describes the shape of a desired product with a geometric model constructed graphically on a cathode-ray tube (CRT) screen of the CAD system. The computer then converts this pictorial representation into a mathamatical model stored in the computer data base for later use. The model may be recalled and refined by the engineer at any point in the CAD process.

Although inserting a model into a computer does not necessarily involve graphics, the design of the object on a computer screen through interactive graphics is one of the most valuable features of CAD systems. Interactive graphics permit the user to communicate easily with the computer through pictures instead of numbers and programs. Interactive graphics allow the user not only to define, store, and manipulate computer pictures, but also to influence the picture as it is being presented. Thus the user interacts with the picture in real time and can study the object by rotating it on the computer screen, separating it into segments and enlarging a specific portion of the object in order to observe it in detail.

CAD systems allow two forms of user interaction: conversational commands and a menu-driven construction mode, in which the user selects a program or a command from a list presented on the computer screen. In practice, interactive graphics has become a prerequisite of every CAD system.

The graphics displays may be of three types: storage tube, raster, and vector refresh. The *storage tube* consists of a CRT with a metal grid located at the rear of a phosphoric screen. The CRT's electron beam strikes the grid at certain points and charges it, thereby causing the phosphor coating at those points to glow. These points on the grid remain charged until the whole grid is reset, and therefore the picture remains on the screen until this reset instruction. Note that the storage tube does not require external memory, since the screen itself stores the picture.

The storage display gives a stable, high-resolution, flicker-free graphic image but is available only in green phosphor with a low-intensity image. This form of display provides the cheapest existing high-resolution display and is available with up to 4096 × 4096 addressable points. However, if elements of the drawing are deleted, they remain visible until the whole screen is rewritten. This feature prohibits the use of the storage tube for animation.

The *raster* gives a line diagram on the screen that is similar to that of a television receiver. The picture is produced by scanning the entire screen with an electronic beam in parallel lines, and modulating the intensity of the beam at each point. Similar to television, the raster gives a high-intensity image which may be viewed in normal room lighting, but the resolution is less than that provided by a storage display. These displays may be obtained with up to 1280 × 1024 displayable points (denoted as pixels), but lower-resolution screens are more typical. The raster image is refreshed 50 or 60 times a second in order to remove any flicker. Both monochrome and full color can be obtained at relatively low cost.

Most current vendors of larger systems are offering workstations with raster displays. One of the greatest disadvantages of the raster display is that lines at a small angle to the horizontal are displayed as a series of steps depending on the screen resolution. However, in practice this is not a serious disadvantage, since it does not affect any plotting. The raster display requires external memory, denoted as the frame buffer, to store the intensity of each pixel. In a monochromatic terminal (two intensity levels) each pixel requires one bit, so a 1024 × 768 display needs a memory of 786,432 bits or about 100 kbytes. If 16 colors or gray levels are added, the storage memory must be four times larger.

The use of raster terminals will probably increase most rapidly because of their low cost and their ability to display a wide range of colors and shades, thereby enabling the display of surfaces rather than curves.

The *vector refresh* display is based on a random-access CRT but uses external memory. The information is displayed as a collection of vectors stored in this memory (the display file). The CRT's electron beam strikes the screen for a short time period only at points corresponding to the display file. For the picture to be maintained, it must be retraced before the eye can detect any decay in glow, namely at a frequency below 50 Hz. Thus the graphic terminal must be refreshed at least 50 times a second if a flicker-free image is to be obtained. This requires intensive processing, and as a result these displays are more expensive than the other two types. There is also a limitation in the maximum number of vectors which can be displayed without flicker (5000 to 50,000) and this can be a severe limitation on some highly complex drawings such as those which occur in mapping (Krouse, 1980). The advantage of the vector refresh display is that vectors can be instantly deleted or added, which makes it a valuable tool for graphical kinematic simulations and display animations of robot manipulators.

There are two basic methods to represent three-dimensional models: the wire-frame and the solid model. Most modeling is done with wire frames that represent the part contour with interconnected line elements. Figures 7-8 and 10-9 represent wire-frame models. These models are using data representation based on point, lines, arcs, or other curves, often stored with view-dependent coordinates.

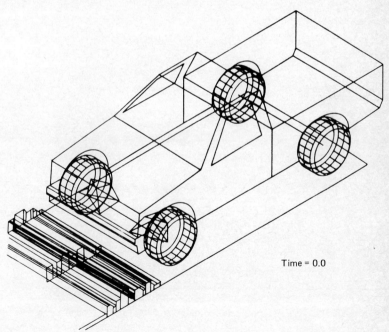

Time = 0.0

FIG. 10-9 CAD simulation of a truck running over a bump. (Mechanical Dynamics.)

One of the most difficult problems in CAD is the elimination of hidden lines. When wire-frame diagrams are used, the computer defines the object without regard to one's perspective. Therefore it will display all the object's contours, regardless of whether they are located on the side facing the viewer or on the back, which normally the eye cannot see.

The wire-frame approach imposes several limitations on the ease with which design data can be passed onto other processes such as finite-element analysis (FEA) and NC. For example, in the stage of NC tape verification, with wire-frame models it is impossible to describe the changing status of a workpiece during machining. It also eliminates the possibility of automatically calculating mechanical properties and checking interference for all but the simplest of objects.

These limitations are caused by the fact that processes such as FEA and NC machining require a complete geometric object description which is capable of characterizing points as being inside, outside, or on

FIG. 10-10 A solid modeling picture is displayed on a graphics terminal. (The picture was taken at Rensselaer Polytechnic Institute.)

the surface of an object. This information is not normally present in the representation produced by wire-frame design systems.

An advanced technique that overcomes this problem is solid modeling (SM). One SM method, denoted as constructive solid geometry (CSG), forms models with building blocks of elementary solid shapes called *primitives*. Primitives are volume elements such as boxes, cylinders, spheres, cones, and ellipsoids. The objects are defined as arbitrary boolean combinations of primitives; one object can be subtracted from another, two objects can be joined (OR), and the intersection of two objects can be found (AND). Subtraction of a cylinder, for example, will create a hole in the displayed part. Figure 10-10 represents an SM picture produced by the geometric modeling technique.

Many commercial CAD programs are of the wire-frame type where a Z dimension has been added to a basic two-dimensional system. Some systems are offering a compromise between true SM and wire-frame models. In these systems surfaces between the wires of the wire frame are defined. A true SM system would permit automatic generation of sections, volume calculations, hidden-line removal, and interference checking.

One of the first SM systems was the SynthaVision program, by which Fig. 7-7 was produced. Other available three-dimensional SM programs include PADL, TIPS, MEDUSA, ROMULUS, EUCLID, CATIA, and COMPAC. The dimensions of the model are stored in the geometric data base and are used in the product analysis and later on in its fabrication and inspection.

Analysis.   After the geometric model is created, some CAD systems allow the user to move directly to analysis. The user may have the computer calculate volume, surface area, moment of inertia, or center of gravity of a part. Other programs allow the use of a materials data base for the calculation of various features of the product, such as strength, stiffness, and weight. The designer can then simulate tests of the product under various environmental conditions, such as temperature changes or different mechanical stresses.

The most powerful method of analyzing a structure on a computer is probably the finite-element method. By this approach the object to be analyzed is broken down into a network consisting of small elements, each of which has stress and deflection characteristics. Graphics displays are used (see Fig. 10-11) for constructing and interpreting the finite-element models. The computer can predict stresses or vibration characteristics which result when the analytical model is subjected to simulated loads.

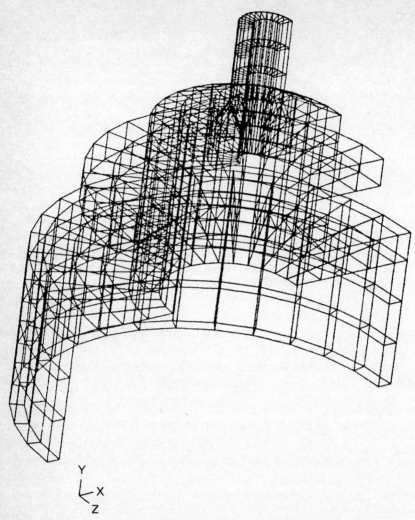

FIG. 10-11 Finite-element model created with the UNISTRUC system. (Control Data Israel.)

The analysis requires the simultaneous solution of many equations, a task which is performed by the computer. Such analysis requires tremendous computational power on large mainframe computers. Once a part is modeled, the user may specify the load parameters and perform an analysis by such programs as NASTRAN, STRUDL, ANSYS, SAP, or UNISTRUC.

Output data, for example, may be displayed showing a deflected shape superimposed over the original model, with deflections exaggerated for evaluation. Or a color-coded model may show areas of high stress. Deflection-mode shapes may even be animated to show deflection during operation.

If the FEA shows too much deflection or other unwanted characteristics, the geometric model may be modified and reanalyzed. Thus, the designer can see how the structure will behave before it is actually built and can modify it without building costly physical models and prototypes.

Another aspect of CAD analysis concerns the combination of analytical and experimental data to create a total system model (Krouse, 1980). In such a system, characteristics of elastic components such as shock absorbers and isolation mounts are determined by testing, while rigid structures are analyzed with finite-element techniques. The data are combined to create a system model, which can then be analyzed with data representing loads. In an automotive application, for example, input data may simulate wheel unbalance, braking, turning, or tire impact with a curb as shown in Fig. 10-9. The computer program responds by predicting the response of the total vehicle to these conditions.

Kinematics.   Many CAD systems have kinematics programs for displaying or animating the motion of robot manipulation and other mechanisms such as four-bar linkages. The kinematics analysis can ensure that moving compounds do not impact other parts of the structure or on its environment. This analysis requires the use of a solid modeler. Computer displays that allow dynamic motion require that the picture be redrawn (refreshed) at least 30 times each second in order to avoid flicker and allow the picture to change smoothly. The design of an automobile, for example, can be facilitated by using kinematics programs. Figure 10-9 shows a graphics simulation of a truck running over an antisymmetric bump in the road. The truck and its suspension are simulated in detail, including geometric, inertial, stiffness, bushing, and damping effects. Figure 10-9 displays the truck at time $t = 0$, but other graphs showing the vehicle crossing the bump can be displayed. Such graphics tests can save the expense involved in producing a prototype of the designed object.

Kinematically, the robot is viewed as a chain of rigid links interconnected by simple, one-degree-of-freedom joints, such as revolute or sliding contacts. Complex joints, if they exist, can be treated as a collection of concurrent simple joints. The assumption of rigid links is made to simplify initial development of the model, but elasticity of members must be included ultimately for completeness. The kinematic solution of a given robot involves the description of position, velocity, and acceleration of various links of the arm without any consideration of loads.

In robotics, the kinematics programs have a twofold usage: design of manipulators and use of graphics as a motion simulator in writing and verification of task programs (see Sec. 7.4). An example of a design of a robot manipulator with a CAD system is given in (Klosterman, Ard, and Klahs, 1982). This example discusses the design process from the initial layout through structural analysis, assembly of the arm, and final examination of the trajectories in space with a kinematics analysis program.

Hundreds of mechanism-designed programs have been developed at universities to solve very specialized problems, but only a handful have been refined sufficiently to be commercially practical. Among these are ADAMS and DRAM developed by M. Chace of the University of Michigan, IMP developed by J. Uicker of the University of Wisconsin, and KINSYN developed by R. Kaufman of George Washington University (Krouse, 1980).

Drafting. With automated drafting, detailed engineering drawings may be produced on a plotter from the data base. In addition, most drafting systems have automatic scaling and dimensioning features.

10.4.2 Computer-Aided Manufacturing

CAM technology is concerned with three main areas: numerical control, process planning, and robotics.

Numerical Control. NC is probably the most mature of CAM technologies. NC and CNC refer to the technique of controlling machine tools with coded instructions which are written with the aid of part programming languages such as the *Automatically Programmed Tool* (APT) language and stored on punched tapes (Koren, 1977, 1983).

Conventionally, the geometric data for part programs are taken from engineering drawings, and the machining conditions are taken from handbooks, catalogs, etc. Based upon this information, the programmer writes the part program. The programmer divides the part contours into simple geometric forms and subsequently writes tool motion commands to cause the machine tool to cut the desired part. The complete part program must then be tested on the machine and debugged several times before the part is machined properly. These iterations significantly increase the cost of machining a part.

CAD/CAM systems make the creation and verification of NC and CNC part programs much more efficient. The source of the geometric data is not human-made part drawings, but the geometric data base of the CAD/CAM system. The elimination of humans from the process of writing geometric dimensions on drawings in the design department and then copying them into part programs in the manufacturing department removes one of the main sources of errors in part programs. The use of a machining data base to obtain cutting conditions (e.g., cutting speed and feed) guarantees more efficient cutting, namely, high productivity, while avoiding tool breakage and reducing tool wear. The machining data base contains data about workpiece material properties, cutting tool geometry and codes, cutting fluids, and machinability data. This data base provides the ability to store vast amounts of data that are necessary to determine cutting conditions as well as to quickly search and retrieve the required data. Finally the CAD/CAM system provides programs for part programs verification. By checking the cutting path with computer graphic simulation, less machine tool time is spent verifying the part programs.

Process Planning. Whereas CNC is concerned with controlling an individual machine tool, process planning considers the detailed sequence of production in the entire factory, including all machine tools and robot systems.

Process planning retrieves information from the machining data base regarding the cutting conditions (in order to predict the cutting time on each machine) and machine tool data, such as horsepower, table size, and available speeds and feeds. Based upon this information, the work can be shared among the machines or the manufacturing cells.

An important element in process planning systems is a concept called *group technology*. Group technology is a method of coding and grouping parts on the basis of their similarities in function or structure or in the ways they are produced. Application of group technology can enable a company to make the production of parts and their movement in the plant more efficient. In one method, robots and CNC machine tools are arranged in manufacturing cells and are able to process a family of similar parts assigned by group technology.

Robotics. Robots are used in the factory to perform a variety of applications, such as material handling, loading and unloading machine tools, assembly, and spot welding (see Chap. 6). Most robots are presently programmed by a manual method, using a teach box to move the robot from one point to the next while recording the points (see Sec. 7.1). This is, however, a very tedious and time-consuming teaching process, and therefore off-line task programming languages are rapidly being developed (see Sec. 7.3).

In future robotic languages, the geometric data required for the program will be received from the CAD/CAM common data base. For example, the designer of an automobile will specify spot-welded points on the car body during the CAD process. These points can be stored in the CAD/CAM geometric data base. The programmer of the welding robot will later use the CAD/CAM data base as a source for obtaining the weld points for the task program. This program, when completed, will be transferred from the CAD/CAM computer through fiber-optic cables into the robot computer for execution. Similarly, task programming of robots in assembly and machine tool loading will use the common geometric data base to obtain information about grip points and trajectories needed in their programs.

The CAD/CAM graphics system and the kinematics programs can be used to solve problems in robot motion. The robot and its environment can be displayed to study the possible collision-free trajectories. Graphic display can aid in the design of the robot workspace from safety viewpoints, displaying the robot in different working states. Note that using the robot's work envelope provided by the manufacturer for safety design is inadequate, since it contains only the arm motion, and does not take into account the wrist and the end effector. Finally, task program verification can be executed with CAD/CAM systems and save debugging time on the robot itself.

In recent years CAD/CAM technology has improved industry productivity and provided a significant step toward the design of the factory of the future.

10.5 THE FACTORY OF THE FUTURE

The concept of the factory of the future has been developed in response to a world trend to batch production, a trend which is driven by the competitive economy in the western countries. Competitive economy means looking for new manufacturing technologies to reduce the price of products. Competitive economy also means more varieties of products and frequent introduction of new products, and it results in lower order quantities. In this sense, the age of mass production is gone and the era of flexible production is being started.

The requirement for flexible production systems dictates the specifications of the factory of the future:

Rapid introduction of new products

Manufacturing of small quantities at competitive production costs

Quick modifications in products of similar function

Consistent quality control

Ability to produce a variety of products

Ability to produce a basic product with customer-requested special modification

The core of the factory which meets these specifications is the CIM system, which was illustrated in Fig. 10-1. The CAD/CAM process shortens the time between the concept point of a new product and its manufacture; the FMS can produce the new product by loading a new program into its supervisory computer; the automatic assembly lines can accommodate the assembly problems of a variety of products with customer-tailored modifications; and automatic inspection maintains the high quality. All this is achieved with few workers on the shop floor, where only material-handling systems, automatic controls, and industrial robots are performing, under limited and remote human monitoring. Raw materials will be entering at one side and finished products will be coming out of the other end.

10.5.1 Factory Management

Studies indicate that in batch manufacturing a part spends about 5 percent of the total time being machined and 95 percent waiting and moving (Cook, 1975; Kegg & Carter, 1982; Koren, 1983). The breakthrough in productivity in the factory of the future will come mainly by cutting down substantially the 95 percent waiting time. This reduction will be achieved by using a computer for managing the factory.

The managing computer is the main tool in resource planning, scheduling, inventory management, and material requirements planning (Koren, 1983). This computer also controls the production flow through the manufacturing and assembly lines and must therefore be interfaced with the supervisory computer of the FMS and the automatic assembly plant.

Manufacturing in the factory of the future is characterized by small quantities of products which are frequently changed. The managing computer must plan a specific sequence of machining operations for each batch. Everything must be ready on time. Too much too soon creates excess capital tied up in in-process inventory. If parts arrive late, the whole operation can be held up, and all the other elements which arrived on time become excess inventory. This high level of coordination can be achieved only with computer technology.

In order to handle the scheduling automatically, the managing computer must have access to the geometric and machining data bases of the CAD/CAM system (see Fig. 10-8). In order to assign the resources and prepare the material requirements planning, access to the materials data base is needed. On the other hand, the managing computer coordinates the operations of the FMS and the assembly-line computers. This eventually leads to interfacing the various computers throughout the factory to a single system, which is the concept of CIM.

10.5.2 The Japanese Unmanned Factory

One of the most ambitious programs for CIM systems has been that known as Methodology for Unmanned Manufacturing (MUM) in Japan. This was begun in 1972 and was carried on cooperatively by a consortium of Japanese government agencies, technical societies, universities, trade associations, and industrial companies. It produced a proposal and plan for developing and building a prototype unmanned factory for production of machine parts and components with $100 million of governmental funding. In this factory, a work crew of about 10 persons, instead of the usual 700, are to be employed. The proposal and plan were completed at the end of 1975 and submitted to the Ministry for International Trade and Industry (MITI) for funding. It failed to receive funding because it was considered to be too idealized a plan and because of political pressures from two other groups, one pushing for a major program of laser machining and one for a major program on work preparation in manufacturing (Merchant, 1980).

The result was a new national project known as Flexible Manufacturing System Complex (FMSC) Provided with Laser (Kimura et al., 1984; Merchant, 1980; Weill, 1983). The project, initiated in 1977, combines aspects of the laser machining and work preparation programs with that for the unmanned factory. Its objective is to establish the technique required for flexible and quick manufacture of complete products and parts of small batch production. The former MUM plan is now considered to be the idealized plan which will provide general guidelines for work in this area for about the next 20 years, while the new program is considered to be the practical plan. Three government laboratories and 20 manufacturing companies are participating in it, and this work is being guided by the Agency of Industrial Science and Technology of the MITI.

The construction of a small FMSC test plant was in its final stage in 1984 (Kimura, 1984). The plant, located in Tsukuba city, is about 720 square meters in size. It consists of four sections: metal forming, machining, assembly, and inspection, with laser processing employed in the first two sections. It is so designed that the FMSC can manufacture all products without changing the common process.

Entering the plant, the raw material is prepared by laser cutting, then formed by a forging process, and put in final shape by the machining process section. Laser heat treatment can be applied if required. Further on, the product is automatically assembled, and at last automatically inspected. Special features of the system comprise the higher-power (20 kW) CO_2 gas laser for material cutting, a medium power YAG laser for heat treatment, complex machining mechanism, and complex assembly process. The complex machining mechanism is a hybrid of machining devices designed for high flexibility and high efficiency. Palletizing and chucking techniques have been specially developed, and provision for automatic diagnosis is ensured. The complex assembly unit will have several powerful functions such as automatic change of assembly tools, detection of part posture, and change of holding position. It is an original development of Toyoda Machine Works for automatic assembly of medium-size parts. The product inspection process will consist of three-dimensional coordinate-measuring machines and also of general performance tests carried out by simulation of actual operating conditions.

By 1985 (the year of termination of the project), the prototype plant is expected to employ about three or four operating workers, plus a somewhat larger support crew than that envisioned for the MUM factory. The program has been granted about $50 million funding by MITI for the 8-year period. Commercialization of results will begin in the late 1980s (Honda et al., 1979; Merchant, 1980; Weill, 1983.)

10.5.3 The Future of the Factory

In the seventies, industries were beginning to emigrate from the United States and Europe to countries where labor was cheaper and more disciplined, thereby leaving the jobless behind and creating an unemployment problem. Factories could survive in the United States only by renewing their technological base and utilizing new strategies of management and production. That is exactly what the concept of the factory of the future offers: a revolutionary change in manufacturing techniques founded on unprecedented involvement of robot systems and computer technology in factory production methods and management. The only way for the economy to recover is to realize "that henceforth our human activities must move away from traditional physical types of work toward endeavors that are the creative fruit of the mind" (Pisar, 1979), while the traditional physical work will be performed in factories by robot systems.

There are industrial leaders who perceive the robot only as a programmable machine that can replace workers. But the key issue is not putting a robot in a shop to do the work of two persons. The key issue is how best to use the robot as a tool in the factory of the future.

The elements of the factory of the future, CAD/CAM, FMSs, robotized assembly, and automatic inspection, are beginning to be used in industry today. Their eventual integration through a computer network will provide the breakthrough to a complete CIM system and the factory of the future.

The size of the factory of the future is likely to be small, not more than several hundred people per plant. But for the manufacture of some types of products, probably fewer than a dozen workers will be employed in the plant. Therefore the present trend of reduction in the number of blue-collar workers in factories is likely to continue as robots and FMSs are installed. Some of the displaced workers will have to be retrained to exploit the new technology, while others might lose their jobs.

The aspiration toward the factory of the future is driven by the competitive economy in industrialized countries and supported by available computer technology and robot systems. It seems that the efforts in developing CIM systems in the United States, Japan, and Europe will make the factory of the future more than an illusion or a dream; it will become a reality in the near future.

REFERENCES

Allan R.: "The Microcomputer Invades the Production Line," *IEEE Spectrum*, vol. 16, January 1979, pp. 53–57.

Barash, M. M.: "Computer Integrated Manufacturing Systems," *Towards the Factory of the Future*, ASME, vol. PED-1, November 1980, pp. 37–50.

Cook, N. H.: "Computer-Managed Parts Manufacture," *Sci. Am.*, vol. 232, February 1975, pp. 22–29.

Drozda, T. J.: "Jumping on the FMS Bandwagon," *Manuf. Eng.*, vol. 90, no. 3, March 1983, p. 4.

Groover, M. P.: *Automation, Production Systems, and Computer Aided Manufacturing*, Prentice-Hall, Inc., Englewood Cliffs, N.J., 1980.

Harrow, P.: "The Selection of CAD Systems," *Electron. Power*, January 1983, pp. 63–68.

Honda, F., et al.: "Flexible Manufacturing System Complex Provided with Laser—A National R&D Program of Japan," *Information Control Problems in Manufacturing Technology*, Pergamon Press, Oxford and New York, 1979, pp. 7–11.

Iwata, K.: "Flexible Manufacturing Systems in Japan," *Proc. UNESCO-CIRP Sem. Manuf. Tech.*, Singapore, November 1982.

Kegg, R. L., and C. F. Carter: "The Batch Manufacturing Factory of the Future," *Robots VI*, paper M. 582–592, April 1982.

Kimora, K.: "Tsukuba FMSC test plant," *The 5th Int. Conf. on Production Eng.*, Tokyo, July 1984, pp. 20–27.

Kimura, M., et al.: "Flexible Manufacturing System Complex Provided with Laser; Concept of the System," *The 5th Int. Conf. on Production Eng.*, Tokyo, July 1984, pp. 12–19.

Klosterman, A. L., R. Ard, and J. Klahs: "Geometric Modeling Speeds Design of Mechanical Assemblies," *Comp. Tech. Rev.*, Winter 1982, pp. 103–110.

Kops, L.: "The Factory of the Future—Technology and Management," *Towards the Factory of the Future*, ASME, vol. PED-1, November 1980, pp. 109–115.

Koren, Y.: "Computer-Based Machine Tool Control," *IEEE Spectrum*, vol. 14, March 1977, pp. 80–84.

Koren, Y.: *Computer Control of Manufacturing Systems*, McGraw-Hill Book Co., New York, 1983.

Krouse, J. K.: "CAD/CAM, Bridging the Gap from Design to Production," *Mach. Des.*, June 12, 1980, pp. 117–126.

Merchant, M. E.: "The Factory of the Future—Technological Aspects," *Towards the Factory of the Future*, ASME vol. PED-1, November 1980, pp. 71–82.

Miller, J. K., D. Starks, M. Hastings, and D. Anderson: "Geometric Modeling Moves CAD," *Ind. Res. Dev.*, January 1983, pp. 80–83.

Pisar. S.: *Of Blood and Hope*, Little, Brown and Co., Boston, Mass., 1980, p. 291.

Stauffer, R. N.: "Flexible Manufacturing System—Bendix Builds a Big One," *Manuf. Eng.*, vol. 87, August 1981, pp. 92–93.

Stauffer, R. M.: "Automating for Greater Gains in Productivity." *Manuf. Eng.*, vol. 87, November 1981, pp. 58–59.

Weill, R.: *Study of Advanced Factory Automation in Japan*, Technion—Israel Institute of Technology, April 1983.

APPENDIX A

Glossary of Robotics Terms

This appendix contains terms used in robotics and related areas, such as vision, CAD/CAM, and numerical control.

Adaptive robot: An intelligent robot that automatically adjusts its task to the changing conditions in the environment. It relies heavily on sensors and artificial intelligence to determine its motions.

AL: A robot-programming language developed at Stanford University, commercially offered by Robot Technology Inc. It provides control of multiple arms in cooperative motion. Runs on PDP-11/45 with 128-kbyte memory.

AML: A manufacturing language, developed by IBM to use with IBM's robot systems. AML runs on IBM series-1 computer with a minimum of 192-kbyte memory. A subset of AML runs on the IBM personal computer, to be used with IBM's 7535 arm.

Arm: The main frame of the robot manipulator; usually contains three axes of motion, enabling movement in three degrees of freedom.

Artificial intelligence: The ability of a machine to respond to a new situation and to solve problems without human interference, or the ability of a machine to learn from its experience.

Backlash: Free play in a gear system in terms of link motion.

Bang-bang robot: See Pick-and-place robot.

Basic resolution unit (BRU): The smallest incremental change in axial position that the feedback device can sense; it is usually equal to the smallest allowable position increment in robot task programs.

Batch manufacture: The production of parts in discrete runs or sets; usually refers to quantities of less than 100,000 parts annually.

Binary digit (bit): In the binary numbering system, only two marks (0 and 1) are used. Each of these marks is called a binary digit, or bit. For example, The decimal number 37 converted to a binary number becomes 100101, and we say that it is made up of six binary digits or bits.

Binary image: A black-and-white image represented in memory as zeros and ones. Images appear as silhouettes on the video display monitor.

Bit map: Digitized representation of a display image as a pattern of bits, where each bit maps to one or more pixels.

Blob: Any group of connected pixels in a binary image. A generic term including both "objects" and "holes."

Buffer: Storage area which receives and subsequently releases transient data.

CADAM: CAD/CAM software system developed by Lockheed (California) and now marketed by IBM and CADAM Inc.

CAM-I: Computer-Aided Manufacturing-International Inc. was formed in 1972 to advance CAM technology. Over a hundred industrial companies, educational institutions, and government agencies in North America, Europe, and Japan pooled resources to form this nonprofit organization. Basically, it provides a convenient conduit for transferring information between companies using computers in design and manufacturing, and users and vendors of numerical control equipment and robot systems. The mailing address is

Computer-Aided Manufacturing-International, Inc.
Suite 1107
611 Ryan Plaza Drive
Arlington, Texas 76011

Cartesian coordinate robot: A robot whose arm's axes of motion lie along perpendicular straight lines.

Cathode-ray tube (CRT): A vacuum tube in which a stream of electrons can be focused on a fluorescent screen, producing lighted traces (e.g., television picture tube).

Closed loop: An automatic control system which uses feedback to measure the effects of the control action and constantly compare them with the desired performance in order to perform automatic corrections.

Computer-aided design (CAD): The use of a computer to assist in the design of a product. Typically, the product's shape can be displayed on the computer screen.

Computer-aided manufacturing (CAM): The use of a computer to assist in the manufacturing of a product. The computer might be used in off-line mode, such as in scheduling or part programming, or in on-line mode, as in the control of machine tools in CNC systems.

Computer numerical control (CNC): A numerical control (NC) system for a single machine tool, including a dedicated minicomputer or micro-computer to perform the basic NC functions.

Continuous-path: A family of robots and NC machines in which the path of the tool while traveling from one point to the next affects the work. The path is controlled by the coordinated motion of the manipulator joints or the machine axes.

Contouring systems: Manufacturing systems with continuous-path capability.

Coordinated axis controlled robot: A point-to-point robot wherein the axes of motion reach the target point simultaneously. All continuous-path robots are coordinated axis controlled.

Cylindrical coordinate robot: A robot whose arm's axes of motion lie along cylindrical coordinates.

Data tablet: Flat-surfaced graphic input device used with a stylus for inking and cursor movement.

Degrees of freedom: The number of independent motions in which the end effector can move, defined by the number of axes of motion of the manipulator.

DIAL: Draper *i*ndustrial *a*ssembly *l*anguage. A robot language developed at Charles Stark Draper Laboratory for assembly tasks, utilizing a force sensor.

Digital: Denotes a discrete state of being, such as the presence or absence of a quantity.

Digitization: The process of converting an analog signal into digital values.

Digitizer: A data tablet that generates coordinate data.

Direct drive: The joint shaft is directly coupled to the rotor of the drive motor without any gearing or other transmission mechanisms.

Direct-view storage tube (DVST): A type of CRT with extremely long phosphor persistence.

Distal: Away from the base, toward the end effector.

Dynamic behavior: Describes how a system performs with respect to time.

Edge: A distinguishable change in pixel values between two regions. Edges correspond to changes in brightness from a discontinuity in surface reflection, or illumination. The edge of an object can be defined by thresholding, i.e., defining all pixels above a threshold value as the object and all at that value or below as background.

End effector: The robot's tool, such as a gripper or a welding gun, attached to the wrist in order to perform the robot's task.

Environment: All the conditions and circumstances in the surrounding of the robot which might affect its operation.

Flexibility: The ease with which a machine or a robot can adjust itself to changes in manufacturing tasks. (Antonym: hard automation)

Flicker: Dimming of a CRT display prior to each refresh, caused when refresh rate falls below the phosphor persistence.

Flip-flop: A device or circuit with two stable states. The circuit remains in either state until a signal causes it to change.

Forearm: The link which is connected to the wrist.

Gain: Amount of increase in a signal as it passes through a control system or a specific control element. If a signal gets smaller, it is said to be attenuated. Gain can also mean the sensitivity of a device to changes.

Gantry robot: A robot suspended from a bridgelike framework.

Gray level: A quantitized measurement of pixel brightness.

Gripper: A device for grasping or holding, attached to the free end of the last manipulator link; also called the robot's hand.

Hard automation: Fixed (not flexible) machinery performing repetitive work of mass-produced parts; cannot be adapted (e.g., programmed) to perform variable-type work.

Hardware: The electronic and mechanical structure of a computer or a robot controller.

HELP: A robotic language developed originally by the Italian company DEA and now offered by GE for use with the Allegro robot. HELP can control multiple cartesian arms in assembly tasks.

Hidden lines: Line segments which should not be visible to a viewer of a three-dimensional displayed object because they are behind other surfaces of the object.

ICAM: *I*ntegrated *c*omputer-*a*ided *m*anufacturing is a U.S. Air Force program attempting to develop one master program that will coordinate all the sophisticated design and manufacturing techniques now employed by industry. Goals include development of hardware and software demonstration manufacturing cells in selected aerospace plants. As part of this program, a robot is being used to make sheet-metal parts for the F-16 aircraft. A major ICAM goal is the complete robotized and computerized sheet-metal fabrication center. The mailing address is

ICAM Program Office
Air Force Materials Laboratory
Wright-Patterson Air Force Base, Ohio 45433

Image processing: Providing an image with properties desirable for analysis, such as increased sharpness, less noise, or reduced data. Also, the extraction of specific information from an image, such as the number and size of the objects in it.

Intelligent robot: Robot capable of deciding its own action with sensory functions and cognitive functions.

Interlock: A device (e.g., a microswitch) which prevents a machine from continuing further operations until a certain condition is fulfilled (e.g., pressing the switch).

Interpolator: A computer algorithm or a hardware circuit, which inserts intermediate points between two given end points of a segment or a trajectory, and emits reference commands to the control loops in order to coordinate their motion to obtain a required path, such as a straight line or a circle; interpolators are used in continuous-path CNC and robot systems.

IPAD: The Integrated Program for Aerospace-Vehicle Design is a NASA-sponsored project, underway since 1976, aimed at developing a software program for integrating existing CAD functions and for developing efficient ways to handle the huge amounts of data involved in such systems. The system is being developed by the Boeing Co. under contract to NASA. The mailing address is.

IPAD Project Office
NASA Langley Research Center
Hampton, Virginia 23665

JIRA: The Japanese Industrial Robot Association, the first in the world, established in 1971. Today there are at least 11 similar organizations in other industrialized countries (United States, Great Britain, France, Italy, Germany, etc.)

Joint: The junction of two member links of the robot manipulator. The members are free to move around the joint.

Lead-through: A means of teaching a robot by leading it through the desired path.

Line of sight: A line from a vision sensor attached to a robot's end effector to a distant point toward which the robot must move.

Link: A connecting element in the robot manipulator.

Main frame: The mechanical arm mechanism which consists of a series (usually three) of links and joints.

Manipulator: A mechanical device, consisting of links and joints driven by actuators, and composed of a main frame (arm) and a wrist.

MCL: *M*anufacturing *c*ontrol *l*anguage, was developed under contract to the U.S. Air Force ICAM program, MCL is an extension to the APT language for robotics use.

Menu: A list of program execution options appearing on the display; the user selects the desired option.

Nonservo robot: A robot controlled through the use of mechanical stops and microswitches.

Numerical control system: A system in which actions are controlled by the direct insertion of numerical data. The system must automatically interpret at least some portion of the data. When a computer is employed as an integral part of the system, it is denoted as a computer numerical control (CNC) system.

Off-line operation: The computer operates independently of the time base of the actual inputs.

On-line operation: Operation where input data are fed directly from measuring devices into the computer. Results are obtained in real time. Can also mean the operation of peripheral equipment in conjunction with the central processor of a computer system.

Open loop: A control system in which there is no self-correcting action for misses of the target value, as there is in a closed-loop system.

Overshoot: Occurs when the robot tool exceeds the target point.

Pattern recognition: A technique that classifies images of objects into categories.

Payload: The maximum weight that can be handled by a robot without failure.

Pick-and-place robot: A nonservo robot which can perform a limited number of sequential actions. The feedback devices are usually adjustable limit switches and mechanical stoppers. Also called bang-bang robot.

Pixel (picture element): An individual element of a raster display (or of a digitized image array), represented as a single point with a specified color or intensity level.

Point-to-point (PTP): A controller type or a mode of operation in which the robot moves to a sequence of predetermined points where the end effector performs the task with the robot stationary. The path of the robot while traveling from one point to the next is without any significance.

Proportional control: A control action in which the motion of the output variable is proportional to its error (i.e., to the difference between the desired and actual values).

RAIL: A robot programming language developed by Automatix Co.

Raster: A video monitor, like a television screen, in which the picture is generated by parallel horizontal lines drawn by an electron gun.

Reach: The maximum horizontal distance from the center of the robot base to the end of its wrist.

Repeatability: The positional deviation from the average position achieved by repetitive motion commands from a start point to a target point. The repeatability of a robot is usually smaller than its accuracy.

Resolution: The smallest change in position that the axial feedback device can sense. See also *Basic resolution unit.*

Resolved motion rate control: A method for solving the inverse kinematics of the manipulator, by which the velocity vector of the end effector is used to determine the corresponding joint velocities.

Robot: See definitions in Sec. 1.4.

RPL: A *robot programming language*, was developed at SRI International. It runs on an LSI-11 under the RT-11 operating system. Can be used with SRI Vision Module, PUMA-600 robot, and other devices.

SCARA: An acronym for selective compliance assembly robot arm. A robot with four axes: two planar articulated motions of the arm and two wrist motions. Shown in Fig. 6-15.

Shoulder: The manipulator joint that is located between the base and the upper arm.

Software: The programs (e.g., control, interpolation, kinematics, and diagnostics) for a computer and microprocessors.

Spherical coordinate robot: A robot whose arm motions are defined by spherical coordinates.

Task Program: A program written by the robot user which specifies the trajectory that the robot should move along in order to accomplish a specific task.

Teach pendant: A hand-held control box with which a robot can be moved and manually programmed.

Teleoperator: A remotely controlled manipulator.

Tool center point (TCP): An imaginary point that lies along the last wrist axis at a user-specified distance from the wrist; e.g., the edge of a welding gun or the middle point between the gripper jaws.

Tool coordinate system (TCS): The tool locations expressed in coordinates relative to a frame attached to the tool itself.

Trajectory: A curve in space through which the TCP moves.

Trajectory planning: The determination of the actual trajectory along which the robot's end effector will move, subject to admissible velocity and acceleration constraints.

TROLL: *T*echnion *r*obotics *l*aboratory *l*anguage, a high-level language for implicit programming of robot assembly tasks, developed at the Technion, Israel Institute of Technology.

VAL: A robot programming language developed by Unimation Inc. to use with their Unimate Series 800 and PUMA series robots.

Verify: To check, usually by automatic means, one typing or recording of data against another, in order to minimize the number of human errors in the data transcription.

Via point: A point through which the robot's tool should pass without stopping; via points are programmed in order to move beyond obstacles or to bring the arm into a lower-inertia posture for part of the motion.

Work envelope: A three-dimensional shape that defines the boundaries that the robot manipulator can reach; also known as reach envelope.

Work volume: The amount of space within which the robot can operate; it includes the entire range of points which can be reached by the robot's wrist.

World coordinate system (WCS): End-effector locations expressed in terms of the cartesian coordinates (and orientation angles of the tool) relative to a reference frame fixed in the base of the robot.

Wrist: The part of the manipulator which is located between the arm and the end effector and contains the most distal group of joints affecting the motion of the end effector.

Yaw: In robotics, a wrist angular displacement to the left or right as viewed from along the principal axis of the wrist.

APPENDIX B

Robot Manufacturers and Suppliers in the United States†

1. **Adaptive Intelligence Corp.**
   2944 Scott Blvd.
   Santa Clara, California 95050
   Tel: (408) 970-0777

2. **Advanced Robotics Corp.**
   777 Manor Park Drive
   Columbus, Ohio 43228
   Tel: (614) 870-7778

3. **AKR Robotics Inc.**
   35367 School
   Livonia, Michigan 48150
   Tel: (313) 261-8700

4. **Akrobotics International Inc.**
   100 Lincoln Street
   Akron, Ohio 44308
   Tel: (216) 376-3344

5. **Amatrol, Inc.**
   P.O. Box 2097
   Clarksville, Indiana 47131
   Tel: (812) 282-4666

6. **American Robot Corp.**
   121 Industry Drive
   Pittsburgh, Pennsylvania 15275
   Tel: (412) 787-3000

7. **Anorad Corporation**
   1100 Oser Avenue
   Hauppauge, New York 11788
   Tel: (516) 231-1990

8. **Applied Robotics, Inc.**
   1223 Peoples Avenue
   Troy, New York 12180
   Tel: (518) 783-1818

9. **Armax Robotics, Inc.**
   38700 Grand River Avenue
   Farmington Hills, Michigan 48018
   Tel: (313) 478-9330

10. **Aronson Machine Co.**
    1001 W. Maia St.
    Arcade, New York 14009
    Tel: (716) 492-2400

11. **ASEA, Inc.**
    1176 E. Big Beaver Road
    Troy, Michigan 48084
    Tel: (313) 528-3630

12. **Automated Assemblies Corp.**
    25 School Street
    Clinton, Massachusetts 01510
    Tel: (617) 368-8914

†List compiled in 1984.

13. **Automation Corp.**
23996 Freeway Park Drive
Farmington Hills, Michigan
48024
Tel: (313) 471-0550

14. **Automation Intelligence, Inc.**
1200 West Colonial Drive
Orlando, Florida 32804
Tel: (305) 843-7030

15. **Automatix Inc.**
1000 Technology Park Drive
Billerica, Massachusetts 01821
Tel: (617) 667-7900

16. **Bendix Robotics**
21238 Bridge Street
Southfield, Michigan 48034
Tel: (313) 352-7700

17. **Benerson Corporation**
1319 W. Florida Street
Evansville, Indiana 47710
Tel: (812) 428-2400

18. **Binks Corp.**
9201 W. Belmont Avenue
Franklin Park, Illinois 60131
Tel: (312) 671-3000

19. **Cincinnati Milacron**
Industrial Robot Division
215 S. West Street
Lebanon, Ohio 45036
Tel: (513) 932-4400

20. **Comet Welding Systems**
900 Nicholas Road
Elk Grove Village, Illinois 60007
Tel: (312) 956-8717

21. **Control Automation Inc.**
P.O. Box 2304
Princeton, New Jersey
Tel: (609) 799-6026

22. **Copperweld Robotics Inc.**
1401 E. Fourteen Mile Road
Troy, Michigan 48084
Tel: (313) 585-5972

23. **Cybotech Corp.**
3939 W56th Street
P.O. Box 88514

Indianapolis, Indiana 46208
Tel: (317) 298-5136

24. **Cyclomatic Inc.**
7520 Convey Court
San Diego, California 92111
Tel: (619) 292-7440

25. **Daido Steel Co.**
Suite 342 East,
Pan American Building
200 Park Avenue
New York, New York 10017
Tel: (212) 986-9117

26. **Dependable-Fordath Inc.**
400 S. E. Willimette Street
Sherwood, Oregon 97140

27. **DeVilbiss Co.**
837 Airport Boulevard
Ann Arbor, Michigan 48104
Tel: (313) 668-6765

28. **Durr Industries Inc.**
40600 Plymouth Road
Plymouth, Michigan 48107
Tel: (313) 459-6800

29. **Dynamcac Inc.**
410 Forest Street
Marlboro, Massachusetts
Tel: (617) 485-9260

30. **Emerson Electric Co.**
Commercial Cam
1440 S. Wolf Road
Wheeling, Illinois 60090
Tel: (312) 459-5200

31. **ESAB North America Inc.**
1941 North Parkway
P.O. Box 2286
Ft. Collins, Colorado
Tel: (303) 484-1244

32. **Everett/Charles**
Automation Modules, Inc.
9645 Arrow Route, Suite A
Rancho Cucamonga, California
91730
Tel: (714) 980-1525

33. **EWAB America**
292 W. Palatine Road

Wheeling, Illinois 60090
Tel: (312) 541-5131

34. **Fared Robotic Systems, Inc.**
3860 Revere Street, Suite D
P.O. Box 39268
Denver, Colorado 80239
Tel: (303) 371-5868

35. **Gallaher Enterprises, Inc.**
2110 Cloverdale
Winston-Salem, North Carolina
27108
Tel: (919) 725-8494

36. **GCA Corp.**
3460 Lexington Avenue
St. Paul, Minnesota 55112
Tel: (612) 484-7261

37. **General Electric Corp.**
Automation Systems Div.
1285 Boston Avenue
Bridgeport, Connecticut 06602
Tel: (203) 382-2876

38. **General Numeric Corp.**
390 Kent Avenue
Elk Grove Village, Illinois 60007
Tel: (312) 640-1595

39. **GMF Robotics Corp.**
5600 New King Street
Troy, Michigan 48098
Tel: (313) 641-4100

40. **Graco Robotics, Inc.**
12899 Westmore Avenue
Livonia, Michigan 48150
Tel: (313) 261-3270

41. **Hirata Corp. of America**
8900 Keystone Crossing
Indianapolis, Indiana 46240
Tel: (317) 299-8800

42. **Hitachi America**
6 Pearl Court
Allendale, New Jersey 07401
Tel: (201) 825-8000

43. **Hobart Brothers Co.**
600 W. Main Street
Troy, Ohio 45373
Tel: (513) 339-6011

44. **Hodges Robotics International Group**
3710 North Grand River Avenue
Lansing, Michigan 48906
Tel: (517) 232-7427

45. **IBM, Advanced Manufacturing Systems**
1000 N.W. 51st Street
Boca Raton, Forida 33432
Tel: (305) 998-2000

46. **Ikegai America Corp.**
2246 N. Palmer Drive, STE 108
Schaunburg, Illinois 60195
Tel: (312) 397-3970

47. **Industrial Automates Inc.**
6123 W. Mitchell Street
Milwaukee, Wisconsin 53214
Tel: (414) 327-5656

48. **Intelledex Inc.**
33840 Eastgate Circle
Corvallis, Oregon 97333
Tel: (503) 758-4700

49. **International Robotic Systems**
6119 Brood Street
Columbus, Ohio 43213
Tel: (614) 863-1051

50. **International Robotmation Intelligence**
2281 Las Palmas Drive
Carlsbad, California 92008
Tel: (714) 438-4424

51. **ISI Manufacturing Inc.**
31915 Grosebeck Highway
Fraser, Michigan 48026
Tel: (313) 294-9500

52. **Keller Technology Corp.**
Robotics Automation Systems
2320 Military Road
Tonawanda, New York 14150
Tel: (716) 693-3840

53. **Kohol Systems, Inc.**
980 Senate Street
P.O. Box 1185
Dayton, Ohio 45401
Tel: (513) 439-5200

54. **Komatsu America Corp.**
    Industrial Machinery Division
    35522 Industrial Road
    Livonia, Michigan 48105
    Tel: (313) 464-6390

55. **Kuclike & Soffa Industries**
    507 Prudential Road
    Horsham, Pennsylvania 19044
    Tel: (215) 674-2800

56. **Kuka, Expert Automation Inc.**
    40675 Mound Road
    Sterling Heights, Michigan
    48078
    Tel: (313) 977-0100

57. **F. J. Lamb Co.**
    5663 E. 9 Mile Road
    Warren, Michigan 48091
    Tel: (313) 536-3535

58. **Lamson Corp.**
    P.O. Box 4857
    Syracuse, New York 13221
    Tel: (315) 432-5467

59. **Lehigh Fluid Power Inc.**
    P.O. Box 246
    York Road, Route 202
    Lambertville, New Jersey 08530
    Tel: (609) 397-3487

60. **Machine Intelligence Corp.**
    330 Potero Ave.
    Sunnyvale, California 94086
    Tel: (408) 737-7960

61. **Mack Corp**
    3695 East Industrial Drive
    Flagstaff, Arizona 86001
    Tel: (602) 526-1120

62. **Manca Inc.**
    165 Carver Avenue
    Westwood, New Jersey 07675
    Tel: (201) 666-4100

63. **Microbot Inc.**
    453-H Ravendale Drive
    Mountain View, California
    94043
    Tel: (415) 968-8911

64. **Mitsubishi Heavy Industries Co.**
    277 Park Avenue
    New York, New York 10017
    Tel: (212) 826-2188

65. **Mobot Corp.**
    980 Buenos Avenue
    San Diego, California 92110
    Tel: (714) 275-4300

66. **Modern Machine Works Inc.**
    Machine Tool Division
    5355 S. Kirkwood Avenue
    Cudahy, Wisconsin 53110
    Tel: (414) 744-5900

67. **MTS System Corp.**
    Ind. Systems Division
    14000 Technology Drive
    Minneapolis, Minnesota 55344
    Tel: (612) 937-4000

68. **Nachi American Inc.**
    223 Veterans Bld.
    Carlstadt, New Jersey 07072
    Tel: (201) 935-8620

69. **NEC America**
    277 Park Avenue
    New York, New York 10017
    Tel: (212) 758-1666

70. **Nissan Motor Mfg. Corp.**
    Smyrna, Tennessee 37167
    Tel: (615) 459-1400

71. **Nordson Corp.**
    555 Jackson St.
    Amherst, Ohio 44001
    Tel: (216) 988-9411

72. **Nova Robotics, Inc.**
    265 Prestige Park Road, L
    E. Hartford, Connecticut 06108
    Tel: (203) 528-9861

73. **Pendar/Kencontronics Inc.**
    1821 University Avenue
    St. Paul, Minnesota 55104
    Tel: (612) 645-7717

74. **Pentel of America**
    2715 Columbia Street

Torrance, California 90503
Tel: (213) 320-3851

75. **Pickomatic Systems, Inc.**
37950 Commerce Drive
Sterling Heights, Michigan
48077
Tel: (313) 939-9320

76. **Planet Corp.**
1820 Sunset Avenue
Lansing, Michigan 48917
Tel: (517) 372-5350

77. **Positech Corp.**
114 Rush Lake Road
Laurens, Iowa 50554
Tel: (712) 845-4548

78. **Prab Robots Inc.**
5944 E. Kilgore Road
Kalamazoo, Michigan 49003
Tel: (616) 349-8761

79. **Precision Robots Inc.**
6 Cummings Park
Woburn, Massachusetts 01801
Tel: (617) 938-1338

80. **Productivity Systems Inc.**
21999 Farmington Road
P.O. Box 279
Farmington, Michigan 48024
Tel: (313) 474-5454

81. **Reeves Robotics, Inc.**
Box S
Issaquah, Washington
Tel: (206) 392-1447

82. **Reis Machines**
1150 Davis Road
Elgin, Illinois 60120
Tel: (312) 741-9500

83. **Rhino Robots, Inc.**
2505 S. Neil Avenue
Champaign, Illinois 61820
Tel: (217) 352-8485

84. **Robotgate Systems, Inc.**
750 Stephenson Highway
Suite 300

Troy, Michigan 48084
Tel: (313) 583-9900

85. **Seiko Instruments, USA Inc.**
2990 W. Lomita Boulevard
Torrance, California 90505
Tel: (213) 530-8777

86. **Sharnoa/Computerized
Machine Tools Inc.**
303 Union Street
Three Rivers, Michigan 49093
Tel: (616) 279-5218

87. **Sigma Sales Inc.**
6505C Serrano Avenue
Anaheim Hill, California 92807
Tel: (714) 630-6553

88. **Sormel/Black & Webster**
281 Winter Street
Waltham, Massachusetts
Tel: (617) 890-9100

89. **Sterling Detroit Co.**
261 E. Goldengate Avenue
Detroit, Michigan 48203
Tel: (313) 366-3500

90. **Sterltech**
P.O. Box 23421
Milwaukee, Wisconsin 53223
Tel: (414) 354-0970

91. **TecQuipment, Inc.**
P.O. Box 1024
Acton, Massachusetts
Tel: (617) 263-1767

92. **Thermwood Corp.**
P.O. Box 436
Dale, Indiana 47523
Tel: (812) 937-4476

93. **Tokico America, Inc.**
3888 W. Lomita Boulevard
Suite E
Torrance, California 90505
Tel: (213) 534-3300

94. **Toshiba America, Inc.**
280 Park Avenue
New York, New York 10010
Tel: (212) 682-8416

95. **Towa Corp. of America**
    1711 S. Pennsylvania Avenue
    Morrisville, Pennsylvania 19067
    Tel: (215) 295-8103

96. **Unimation Inc.**
    Shelter Rock Lane
    Danbury, Connecticut 06810
    Tel: (203) 744-1800

97. **United States Robots Inc.**
    650 Park Avenue
    King of Prussia, Pennsylvania
    19406
    Tel: (215) 768-9210

98. **United Technologies,
    Automotive Product Division**
    5200 Auto Club Drive

Dearborn, Michigan 48126
Tel: (313) 593-9600

99. **VSI Automation Assembly Inc.**
    165 Park Street
    Troy, Michigan 48084
    Tel: (313) 588-1255

100. **Westinghouse Electric Corp.**
     400 Media Drive
     Pittsburgh, Pennsylvania 15205
     Tel: (412) 778-4347

101. **Yaskawa Electric America Inc.**
     305 Era Drive
     Northbrook, Illinois 60062
     Tel: (312) 564-0770

APPENDIX C

Robotics Research Organizations in the United States

There are four categories of organizations in which research in robotics is conducted: nonprofit laboratories, government, universites, and private industry.

Two nonprofit laboratories have been performing research in robotics since the seventies: SRI International and Draper Laboratories. In the government category, research has been primarily conducted by the National Bureau of Standards (NBS), the Air Force, NASA, and the Navy.

The number of universities conducting research in robotics is rapidly growing. Two universities, Stanford and MIT, have been conducting research in robotics since the mid-sixties and can be regarded as the pioneers in this field. In the mid-sixties a remote-controlled manipulator was developed for NASA at Case University in Cleveland and was later used in NASA's Space Nuclear Propulsion Program.

During the seventies, research in robotics has been started at other institutions, such as: University of Florida, Georgia Tech, Ohio State, Purdue, University of Rhode Island, UCLA, and University of Wisconsin. However, the big boom in robotics research happened in the years 1980 to 1982. During a two-year period many large robotics research centers have been established around the country, each of them highly supported by industry and government and involving dozens of professors and other university researchers. In this wave we find the robotics centers at Carnegie-Mellon, Lehigh, University of Michigan, Rensselaer Polytechnic Institute, Purdue, and University of Wisconsin. The directory provides a partial list of universities with research programs in robotics as well as nonprofit laboratory and government institutes.

1. **California Institute of Technology**
   Jet Propulsion Laboratory
   Pasadena, California 92115

2. **Carnegie-Mellon University**
   The Robotics Institute
   Schenley Park
   Pittsburgh, Pennsylvania 15213

3. **Charles Stark Draper Laboratory, Inc.**
   555 Technology Square
   Cambridge, Massachusetts 02139

4. **Georgia Institute of Technology**
   School of Mechanical Engineering
   Atlanta, Georgia 30332

5. **Illinois Institute of Technology Research Institute**
   Manufacturing Productivity Center
   10 W. 35th Street
   Chicago, Illinois 60610

6. **Lehigh University**
   Packard Lab 19
   Bethlehem, Pennsylvania 18015

7. **Massachusetts Institute of Technology (MIT)**
   Artificial Intelligence Laboratory
   Cambridge, Massachusetts 02139

8. **National Bureau of Standards (NBS)**
   Building 220, Room A123
   Washington, D.C. 20234

9. **Naval Research Laboratory**
   Code 7505
   Washington, D.C. 20375

10. **Ohio State University**
    Department of Electrical Engineering
    Columbus, Ohio 43210

11. **Pennsylvania State University**
    Department of Industrial and Management Systems
    State College, Pennsylvania 16802

12. **Purdue University**
    The Computer-Integrated Design, Manufacturing, and Automation Center
    West Lafayette, Indiana 47906

13. **Rensselaer Polytechnic Institute**
    Center for Manufacturing, Productivity and Technology Transfer
    Troy, New York 12181

14. **SRI International**
    333 Ravenswood Ave.
    Menlo Park, California 94025

15. **Stanford University**
    Department of Computer Science, AI Lab.
    Stanford, California 94305

16. **United States Air Force**
    ICAM Program, AFWAL/MLTC
    Wright-Patterson AFB, Ohio 45433

17. **University of California—Berkeley**
    Department of Mechanical Engineering
    Berkeley, California 94720

18. **University of Florida**
    Center for Intelligent Machines and Robotics
    Gainsville, Florida 32061

19. **University of Michigan**
    Center for Robotics and Integrated Manufacturing
    Ann Arbor, Michigan 48109

20. **University of Rhode Island**
    College of Engineering
    Kingston, Rhode Island 02881

21. **University of Wisconsin—Madison**
    Center for Advanced Manufacturing and Robotics
    Madison, Wisconsin 53706

APPENDIX D

National Robotics Organizations

Italy

Società Italiana Robotica Industriale (SIRI)
Via Mantegna, 6
Milano

Japan

Japan Industrial Robot Association (JIRA)
Kikaishinko Bldg.
3-5-8 Shibakoen
Minato-Ku
Tokyo

Netherlands

Contactgroep Industriele Robots (CIR)
Landbergstraat 3
2628 CE Delft

Sweden

Swedish Industrial Robot Association (SWIRA)
Box 5506
Storgatan 19
S-14 85 Stockholm

United Kingdom

British Robot Association
28-30 High Street
Kempston, Bedford MK42 7AJ

United States

Robotic Industries Association (RIA)
(formerly known as Robot Institute of America)
One SME Drive
P.O. Box 930
Dearborn, Michigan 48128

Index

343

ABOUT THE AUTHOR

Yoram Koren received his B.Sc. and M.Sc. degrees in electrical engineering and his D.Sc. in mechanical engineering from the Technion, Israel Institute of Technology in Haifa. Currently Professor of Mechanical Engineering and head of the Robotics Laboratory at the same institution, he has served as a consultant to European and American industry, and to the Israeli Air Force. Dr. Koren has taught at McMaster University in Canada, the University of Wisconsin, and at the University of Michigan where he was Goebel Chair Professor and the Director of the Integrated Design and Manufacturing Division. He is the author of over sixty papers and three books. Professor Koren's *Computer Control of Manufacturing Systems* received the SME's 1984 Manufacturing Textbook award.

Robot